LEADERSHIP & MANAGEMENT

in Health & Social Care for NVQ/SVQ Level 4

Joanne Mckibbin and Alix Walton

Heinemann

Heinemann is an imprint of Pearson Education Limited, a company incorporated in England and Wales, having its registered office at Edinburgh Gate, Harlow, Essex, CM20 2JE. Registered company number: 872828

www.heinemann.co.uk

Heinemann is a registered trademark of Pearson Education Limited

First published 2008

12 11 10 09
10 9 8 7 6 5 4 3 2

British Library Cataloguing in Publication Data
A catalogue record for this book is available from the British Library

ISBN 978 0 435 500 20 7

Edited by Sarah Christopher
Typeset by Saxon Graphics Ltd, Derby
Original illustrations © Pearson Education Ltd 2008
Cover illustration © Pearson Education Ltd 2008
Printed in the UK by Scotprint

There are links to relevant websites in this book. In order to ensure that the links are up to date, that the links work, and that the sites are not inadvertently linked to sites that could be considered offensive, we have made the links available on the Heinemann website at www.heinemann.co.uk/hotlinks. When you access the site, the express code is 0207P

Contents

Acknowledgements

Author acknowledgements

We would like to thank all of the health and social care professionals, students and service users who, directly or indirectly, have contributed to our learning and development, enabling us to be in a position to write this book.

Thanks also to the staff at Pearson Publishers.

Thanks also to Lynda Mason for her invaluable advice and support in completing this book. Also thanks to Jane Kellas from Venus Training.

We would also like to thank our families and friends for their support; in particular thanks to Peter and Iris.

Publisher acknowledgements

The publishers would like to thank the following organisations for their kind permission to reproduce material in this book:

Page 26 – Grant Carson, Glasgow Centre for Inclusive Living [diagram]

Page 55 – General Social Care Council

Page 61 – Skills for Care Providing Effective Supervision, cited by Walsall Local Council

Page 80 – University of Victoria, Canada

Page 87-88 – Research commissioned by Skills for Care and conducted by TNS

Page 95-96 – Social Care Institute for Excellence (SCIE)

Page 106 – Stoke-on-Trent City Council

Page 124 – Healthcare Commission

Page 127 – Commission for Social Care Inspector (CSCI)

Page 123, 212–13 – Crown Copyright material reproduced with the permission of the Controller of HMSO

Pages 214-24 – Skills for Care (www.skillsforcare.org.uk); Skills for Care and UK Development Alliance [Standards]

The publishers would like to thank the following for their kind permission to reproduce photos and images:

Page 72 – Pearson Education Ltd/Gareth Boden
Page 113 – Pearson Education Ltd/Jules Selones
Page 26, 88, 102, 160 – Pearson Education Ltd/Lord and Leverett
Page 78 – Pearson Education Ltd/Mind Studio
Page 14 – Paula Solloway/Alamy

About the authors

Joanne McKibbin

Joanne has worked in the care and education sector for 25 years. This has included the management of people and services, and teaching in the further education sector. Joanne has worked with NVQ/SVQ's for over 15 years and holds NVQ/SVQ assessor and internal verifier awards. Joanne is currently an NVQ/SVQ centre manager, workforce development officer and freelance trainer and consultant.

Alix Walton

Alix is a registered Social Worker with over 20 years experience in the health and social care and education sector. This has included working as a Senior Practitioner in Social Services and managing a training centre. Alix has worked with NVQ/SVQ's for over 12 years and holds NVQ/SVQ assessor and internal verifier awards. Alix currently lectures on health and social care programmes in Higher Education and is a freelance trainer and consultant.

Introduction

Welcome to Leadership and Management in Health and Social Care for NVQ/SVQ Level 4. This textbook will aim to support you, not only through this qualification but also throughout your career as a manager.

We hope you find this book a useful resource in your current role and as part of your continuing professional development. While it has been written to cover the areas of practice and knowledge needed to complete the Leadership and Management NVQ/SVQ Level 4 Award it is also relevant for managers completing a wide range of training programmes and qualifications.

It will also be helpful for those people who are interested in becoming managers as it covers many of the skills and knowledge areas that you need to develop in order to become an effective manager of people, resources and settings.

Leadership and Management for Care Services

To complete the Leadership and Management NVQ/SVQ Level 4 Award you will need to show your practice, skills and knowledge across eight units: this is a combination of four mandatory and four optional units. The mandatory units are:

- Manage and develop yourself and your workforce within the care service (A1)
- Lead and manage provision of care services that respects, protects and promotes the rights and responsibilities of people (B1)
- Develop and maintain systems, procedures and practice of care services to manage risks and comply with health and safety requirements (C1)
- Lead and manage effective communication that promotes positive outcomes for people within care services (E1)

It is important that you select optional units that relate to your current role as your practice for each unit will need to be observed in your work setting by a qualified assessor. Your assessor will be allocated by the assessment centre which you are registered with. There is a wide range of optional units to choose from but if you have difficulty selecting appropriate ones you can discuss this with your assessor who will be able to advise you.

You will also compile a portfolio of evidence, which will demonstrate your skills, knowledge, values and understanding. A portfolio is usually a file where you keep all of your evidence and assessment records. Your assessment centre will explain how they want you to organise your portfolio but it is important it is well structured and indexed so that it is clear and easy to use.

Familiarising yourself with the Leadership and Management standards will help you when planning observations of your practice and ensuring your written work contains enough detail to meet the unit criteria.

Traditionally books based on NVQ/SVQ qualifications have taken a unitised approach. This book is significantly different. Rather than chapters based on specific units they are each based on areas and issues of relevance to all managers in the health and social care sector, presenting a more holistic approach to practice and NVQ/SVQ assessment. One topic area or piece of work can cover aspects of several units, as there are common themes, performance criteria and knowledge statements across units.

Within each chapter links to NVQ/SVQ Leadership and Management units have been identified.

Features in this book

There are several features which appear consistently through the book.

- **Activities** – these are exercises for you to complete, or questions for you to think about. These will help test your knowledge and understanding and possibly help you identify areas for development.
- **Case studies** – enable you to explore different scenarios and issues related to management practice. This gives you the opportunity to broaden your knowledge and understanding and helps you to develop strategies to deal with potentially difficult situations. Case studies can be used in your NVQ/SVQ portfolio **but not** as a main assessment method. They can be used to cover some gaps in knowledge where you are having difficulty demonstrating your practice, or they can be used as tool to help you clarify your thought processes, which can assist you when writing reflective accounts. Case study answers are available in appendix 1 at the back of the book, enabling you to check your understanding and the conclusions you have drawn about the situations described in the features throughout the book.
- **Reflect** – the reflect sections help you to consider issues and reflect on areas of practice, skills, values and knowledge. They will also help you to develop reflective accounts which can be used as evidence in your portfolio. In reflective accounts you will link your practice to your knowledge, values and understanding, and cross reference them to relevant NVQ/SVQ units. Chapter 1 gives guidance on how to write reflectively and structure reflective accounts.
- **Viewpoints** – give you the opportunity **to** identify and review theories, legislation, policies and procedures and how these can influence and impact on your practice.
- **NVQ Link** – these show you where the information you have read, or collected, connects with NVQ/SVQ units.
- **Appendices** – these provide additional information that you may find useful both for your qualification and throughout your career as a manager. Appendices included are:
    - Case study answers
    - Useful forms and templates
    - The adult social care manager induction standards from Skills for Care
    - Mapping grids for the NVQ/SVQ and Management Induction standards
- **Hotlinks** - all websites referred to in the book are also available as hotlinks at www.heinmann.co.uk, giving you easy access to a wealth of information from one easy starting point. These hotlinks mean you can look into certain areas in more depth and can keep yourself up to date with legislation and key professional practice. Forms and templates from appendix 2 are also available through www.heinemann.co.uk, giving you the flexibility to download and use these for your own purposes.

Work products

A work product is a piece of evidence that you have generated or developed as part of your work in your setting. Work products should clearly demonstrate that you are working at the right level, in a managerial position and putting aspects of the unit(s) criteria into practice. They could be minutes of a meeting you have been involved in, a report you have written or an appraisal (this could be yours or one you have written for a team member).

Many work products contain confidential information so they will not necessarily be included in your NVQ/SVQ portfolio – your assessor will advise you about the guidance and regulations relevant to the NVQ/SVQ Centre you are registered with. If you are unable to include some work product in your portfolio because of confidentiality issues you can show the work product to your assessor who can write a confidential work product sheet, thus allowing you to still be credited with the skills and knowledge you have demonstrated. Your assessor will clearly highlight the general content of the work product and the skills and practice you have demonstrated, but will omit details such as names and places, to ensure confidentiality.

Quality vs. quantity

To successfully complete your NVQ/SVQ Award you will need to cover all performance criteria and knowledge statements in each unit. Your portfolio should only comprise work that you have written and/or been involved in developing. For instance, if you have written or contributed to the writing of any policies, procedures or information leaflets then you can include them on your file. You will need to outline your involvement and clearly show how they meet relevant performance criteria and knowledge statements. If you have not been involved in writing such documents then they cannot be included in your file, as they are not evidence of **your** practice.

Handouts, reading material or other material collected as part of your NVQ/SVQ programme should also not be included as evidence in your NVQ/SVQ portfolio, as they also do not demonstrate your knowledge or practice. However you may wish to keep these in a separate file so you are able to access them easily if you want to use them to reinforce your knowledge and understanding, and to provide references for your written work. Chapter 1 provides guidance on referencing your work.

Select your evidence wisely. Think as if you are going on holiday. When you pack your suitcase you only take clothes that are appropriate to the trip you are going on, you don't pack your whole wardrobe (however tempting!). This means you have to select your clothes wisely, and you should select evidence for your NVQ/SVQ portfolio in a similar way.

Further reading

Suggestions have been made throughout the book regarding additional reading that you can do to help you develop your knowledge and understanding further. There are also useful web links which will give you access to a wide range of up-to-date material. Law, policy and regulations change regularly within the health and care sector, and part of your continuing professional development is to ensure you remain up to date with these changes, as well as with current research in your area of practice. The suggested reading and websites will help you to do so, both during and after your NVQ/SVQ studies.

Finally, we hope you enjoy using this book and wish you good luck with your management studies!

CHAPTER 1

Reflective practice and writing

'Reflective practice is associated with learning from experience. Experience alone is not sufficient for learning.'
(Boud et al., 1985:7)

'Reflective practice is important to the development of all professionals because it enables us to learn from experience. Although we all learn from experience, more and more experience does not guarantee more and more learning. Thus twenty years of work in social care may not equate to twenty years of learning about social care, but may be only one year repeated twenty times.'
(Beaty, 1997:8)

Introduction

You are likely to already be reflecting on your practice, even if you are doing so unconsciously, but to complete your management award you will need to evidence this process. You will need to show that your practice is underpinned by knowledge and understanding, and that you are able to analyse what you do, how you do it and why you do it, as well as analysing what you have learned from your practice and experiences. In this chapter, you will read about:

- the purpose of reflection
- models of reflection
- reflective accounts: writing effectively
- some tips for studying.

Before reading on, carry out this short activity.

ACTIVITY

Reflect on a significant event

Think about the last time you went to a significant event involving family or friends, such as a birthday, wedding or other celebration.

Now identify one or two family members or friends from the significant event and consider the following:

- What thoughts did you have during or after the event about areas such as how long you have known the person/people involved; the other life events you had shared with them or seen them overcome; how they have changed, physically, emotionally and intellectually?
- Can you identify anything you have learned from the relationship or experiences you have shared?
- How have you changed because of this?

The process you went through during and after the special event, and in considering these questions now, is called reflection. It is a process you may often engage in in your daily life, but it is also important that you reflect in your professional life.

The purpose of reflection

Beaty (1997) stresses that 'we should not rely solely on our natural process of reflecting on experience, but actively seek ways to ensure that reflection itself becomes a habit, ensuring our continuing development…' and in work this is especially true – but why?

Begin by considering the purpose of your own reflections at work, by completing the activity below.

ACTIVITY

Reflecting on your work practice

Using the list below, identify the reasons why you might reflect on your work practice. Tick as many reasons as you feel are relevant to your situation.

Purpose of reflection	Tick those that are relevant to you
To record experience	
To facilitate learning and development	
To support understanding	
To develop knowledge	
To develop critical thinking	
To develop evaluative skills	
To enhance problem-solving skills	
For self-empowerment	
To support behaviour and feelings	
To develop teamwork	
To enhance creativity	
To improve written skills	
To improve communication	
To support project work	
To provide evidence for an NVQ/SVQ or other qualification	
To provide evidence for continued professional registration	
Other reasons: please specify	

(Adapted from McGill and Brockbank, 1998)

Now consider these questions.

- How many of the boxes did you tick?
- Did completing the list highlight that you reflect on a wide range of areas and for different purposes?
- Did it show that this is an area of your practice that you may need to develop further?

Models of reflection

Reflection is not just about the knowledge held by an individual. Reflection 'should challenge the concepts and theories by which you make sense of the knowledge. When you reflect, you don't simply see more; you see differently' (Platzer et al., 1997).

There are a number of models of reflection, and three of the best known are outlined in this chapter. A model is a broad framework, within which a range of theories can be applied; theories help you to understand 'why'. You may use any of the models of reflection explained below, and within that model you can use different theories to help you understand different aspects of the event. For example, you might use Johns' model of reflection (see Figure 1.4) to help you to reflect on a situation when a service user became distressed, and consider theories of loss or attachment to help understand why the service user became distressed.

> **NVQ/SVQ**
>
> **Additional sections have been added to each of the model templates, to allow you to identify the NVQ/SVQ units, elements and knowledge statements evidenced within your reflective accounts.**

Kolb's Learning Cycle

David A. Kolb is Professor of Organizational Behavior at the Weatheread School of Management in Cleveland, USA. Kolb (1984) identified reflection as part of the learning cycle (see Figure 1.2 and Chapter 4), believing that, to enable effective learning and development, reflection needed to be part of the process.

Kolb's Learning Cycle suggests that it is not sufficient to have an experience in order to learn; it is necessary to reflect on the experience, draw on our knowledge base and identify new ideas and concepts, which can then be applied to new situations. Kolb stresses that, without reflection, we would simply continue to repeat our mistakes.

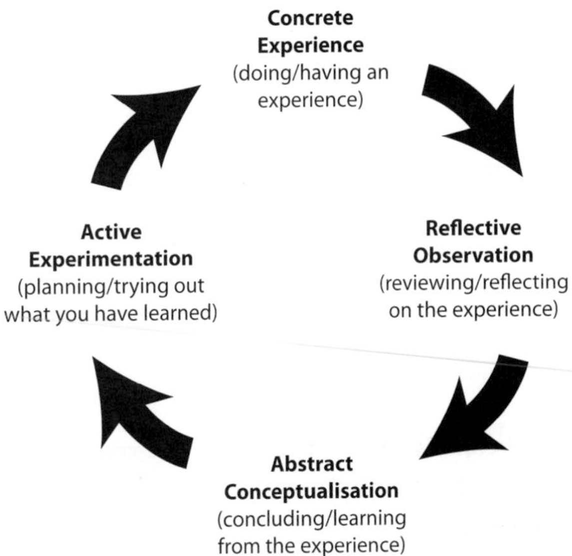

Figure 1.1 Kolb's Learning Cycle

Concrete experience: the event or experience • What happened? • How did you feel during the experience? • How did you react?
Reflective observation: reflecting on the event or experience • What went well? • What did not go well? • Did anything unexpected occur? • How do you feel about the experience now?
Abstract conceptualisation: drawing on your knowledge base • How does theory inform your understanding of the experience? • How does research inform your understanding of the experience? • What conclusions can you draw from this?
Active experimentation: planning for the next experience • What other options did you have? • What might you do differently in the future? • What other strategies or methods might you use in the future?
This reflective account provides evidence for: (list which units/elements/knowledge statements this will meet)

Figure 1.2 Kolb's model of reflection template

Gibbs' Experiential Learning Cycle (1988)

Professor Graham Gibbs is Director of the Oxford Learning Institute. His reflective model is fairly straightforward, breaking the reflective process down into clear stages:

- **description** of the situation
- analysis of **feelings**
- **evaluation** of the experience
- **analysis** to make sense of the experience
- **conclusion**, where other options are considered
- **action plan** for future learning and development.

Description What happened?
Feelings What were you thinking and feeling?
Evaluation What was good and bad about the experience?
Analysis What sense can you make of the situation?
Conclusion What else could you have done?
Action plan If it arose again, what would you do?
This reflective account provides evidence for: (list which units/elements/knowledge statements this will meet)

Figure 1.3 Gibbs' reflective model template

Johns' Structured Reflection (1994)

Chris Johns developed his model to facilitate a process of guided reflection. His focus was on uncovering and making explicit the knowledge that we use in our practice. The model was originally developed in a nursing context, but can be adapted for use in a wide range of work settings and professions. Johns' model divides the reflective process into five key areas.

Description of the experience

Reflection

- What was I trying to achieve?
- Why did I intervene as I did?
- What were the consequences of my actions for:
  - myself?
  - the service user/family?
  - the people I work with?
- How did I feel about this experience when it was happening?
- How did the service user feel about it?
- How do I know how the service user felt about it?

Influencing factors

- What internal factors influenced my decision making?
- What external factors influenced my decision making?
- What sources of knowledge did/should have influenced my decision making?
- How did my actions match my beliefs?
- What factors made me act in an incongruent way?

Evaluation

- Could I have dealt with the situation better?
- What other choices did I have?
- What would be the consequences of these choices?

Learning

- How do I now feel about this experience?
- How have I made sense of this experience in light of past experiences and future practice?
- How has this experience changed my ways of knowing, in terms of:
  - empirics – scientific?
  - ethics – moral knowledge?
  - personal – self-awareness?
  - aesthetics – the art of what we do, our own experiences?

This reflective account provides evidence for: (list which units/elements/knowledge statements this will meet)

Figure 1.4 Johns' model of reflection template

ACTIVITY

Comparing reflective models

Select an event that you have experienced in the last fortnight in your work setting that you would like to reflect on. Use both Gibbs' and Johns' templates to reflect on the event, then consider the following questions.

- Which model do you prefer, and why?
- Would you add or change any headings or questions within the templates?

It is important that you find a structure for reflection that suits your own style when writing reflectively. There are strengths and weakness within all models of reflection. Remember: you do not have to use only one model, or use all the headings that are listed. You can use different models in different circumstances, and change or add headings that enable you to write effectively and meet the criteria needed for assessment purposes.

Moon (2001) stresses a number of key points regarding reflective writing, and particularly emphasises the importance of practising and improving this skill.

- Be aware of the purpose of your reflective writing and state this, if appropriate.
- Reflective writing requires practice and constant standing back from oneself.
- Practise reflecting writing on the same event/incident through different people's viewpoints and disciplines.
- Deepen your reflection/reflective writing by discussing issues with individuals and groups, getting the points of others.
- Always reflect on what you have learned from an incident, and how you would do something differently another time.
- Try to develop your reflective writing to include the ethical, moral, historical and socio-political contexts, where these are relevant.

Reflective accounts: writing effectively

There is no right or wrong way of structuring your reflective accounts, as long as you include sufficient detail to meet the unit criteria you are being assessed against. Each of the models above provides a different format for reflective writing; what is important is that you use a format that enables you to write effectively to complete your qualification. This section provides you with a number of models and checklists to use to evaluate the quality of your reflective writing.

One of the common problems with reflective writing is that it can be overly descriptive and lacking in analysis. Cottrell (2003) provides some useful guidance to highlight the difference between descriptive and critical analytical (reflective) writing (see Figure 1.5).

Bloom's Level of Thinking Processes (1964) also highlights the difference between description and critical analysis and evaluation (see Figure 1.6). When writing reflectively, you should aim for the higher level processes identified here.

Moon (1999) and Winter et al. (1999) offer some helpful assessment criteria for reflective writing. Although these were not devised specifically for NVQ/SVQs, you may find them helpful as checklists to evaluate the development of your own reflective writing.

Descriptive writing	Critical analytical writing	Examples
States what happened	Identifies the significance	Why is a piece of information important? Why have you included it in your account?
States what something is like	Evaluates (judges the value of) strengths and weaknesses	What worked, what didn't work and why?
Gives the story so far	Weighs one piece of information against another	Why is one factor more important than another in experience or area of practice considered?
Explains what a theory says	Shows why something is relevant or suitable	How did theory, e.g. change management theory, inform your practice?
Notes the method used	Indicates whether something is appropriate or suitable	Why did you use a particular method? Did it work or not?
States options	Gives reasons for selecting each option	Why did you select that particular option? Why did you reject other options?
List details	Evaluates the relevant significance of details	Why was a particular factor, e.g. the ethnicity of the service user, significant?
Gives information	Draws conclusions	What worked, what didn't work, what did you learn?

(Adapted from Cottrell, 2003)

Figure 1.5 Differences between descriptive and critical analytical (reflective) writing

Increasing Difficulty ↓

Process	Explanation
Knowledge	Recognition and recall of information – describing events.
Comprehension	Interprets, translates or summarises given information – demonstrating understanding of events and stating event and understanding of it in own words.
Application	Uses information in a situation different from original learning context – how learning is applied in new situations.
Analysis	Separates wholes into parts until relationships are clear – breaks down experiences, identifies connections between elements and can distinguish fact from opinion.
Synthesis	Combines elements to form new entity from the original one – draws on experience and other evidence to provide new insights.
Evaluation	Involves acts of decision making, or judging based on criteria or rationale – makes judgements about the value of elements or ideas.

(Adapted from Bloom, 1964)

Figure 1.6 Bloom's Level of Thinking Processes

Moon (1999) identifies the following criteria with which to assess quality of reflective writing:

- demonstrates an awareness and understanding of the **purpose** of the journal (reflective account), and uses it to guide selection and description of events/issues
- **description** provides adequate focus for further reflection and includes additional ideas
- **reflective thinking** is evident – ability to work with unstructured material, make a theory – practice link, show awareness of different points of view of the event and alternative interpretations, application of theory, testing of new ideas
- **product** – there is a statement of something that has been learned or solved, a sense of moving on.

Winter (1999) identifies the following criteria:

- includes careful, detailed observation of events and situations
- empathises with the standpoint of other people
- notices the various emotional dimensions of events and situations
- addresses the complexities of issues, events and situations
- makes connections between different events and situations, and between specific details and general principles derived from a range of professional knowledge
- demonstrates learning, in response to both professional experience and the process of reflecting upon/writing about it.

ACTIVITY

Reviewing your reflective accounts

Return to the reflective accounts you completed earlier using Gibbs' and Johns' models of reflection (see page 4). Select one of the four models outlined above (Cottrell, Bloom, Moon or Winter) and use it to evaluate your reflective accounts. Identify the strengths and areas for ongoing development.

NVQ/SVQ

The following activity will help you draw together your learning from the chapter so far, and provide a further opportunity for you to practice writing reflectively, as well as to generate some evidence for your NVQ/SVQ.

ACTIVITY

Reflecting on your job role

Write a reflective account on your current job role as a manager. Include the full range of responsibilities you have and the types of skills you need to carry them out. Say **what** you do, **when** you do it, **why** you do it, **how** you do it, **who** you involve and **where** it takes place.

Once you have completed this, focus on one area of your practice that you think you do well. Identify what it is, why you feel you do it well and how you cascade this area of good practice to your team.

Then highlight an area of your practice that you feel you need to develop. Identify why this is an area that you need to develop and any ways in which you can improve this area. Highlight whether you will need to involve anyone else to help you develop this area (and say why), how you will measure your learning and development and how this will benefit your setting.

Reflective practitioners 'recognise ethical dilemmas and conflicts and how they arise. They are more confident about their own values and how to put them into practice; they integrate knowledge, values and skills; reflect on practice and learn from it; are prepared to take risks and moral blame' (Banks, 1995). Writing this reflective account will help you to identify the range of skills and responsibilities you have, and to analyse how you carry them out. As Banks (1995) suggests, a practitioner who can reflect on their practice and learn from it tends to be more confident, which benefits the settings he or she works in.

NVQ/SVQ

Once you have completed the activities, look at the units – both the mandatory units and those optional units you are considering – identifying where and how your practice links to them. This will help you to identify the areas of practice where you may already have evidence to complete the units, and the areas where you will need to develop your knowledge and practice.

Some tips for studying

Before beginning to study it is worth spending some time planning how you will organise your study time and you may find it helpful to start by identifying your learning preferences. By doing so you are more likely to ensure you take control of your study environment and make the most effective use of the time you have available.

TO-DO LIST

NAME: **DATE STARTED:**

Task to complete	Target date	Tick when completed

Figure 1.7 A to-do list

ACTIVITY

Learning Preferences

- Where do you prefer to learn, e.g. home, office, library or other?
- Do you prefer to learn alone or with others?
- Do you prefer to learn by doing practical tasks, thinking for yourself, watching others, reading or another way?
- Do you prefer it to be to quiet or for there to be noise?
- Do you prefer to work at a desk, table or sofa?
- Do you learn best in the morning, afternoon, evening?
- Do you prefer to have a supply of food and drink?
- What other conditions do you prefer, e.g. to be warm, for there to be natural light?

Having answered these questions, identify how you will manage your study environment and time, and any actions needed to prepare for your study, e.g. moving a desk or table to a window or radiator.

(Adapted from: Beverley & Worsley, 2007)

You can encounter a range of difficulties while studying. Being aware of potential problems, how they might affect you and ways you can reduce the effects will help you to achieve your study goals.

REFLECT

Try to identify the potential difficulties you may encounter while working towards your qualification. How can you overcome the difficulties? Who can help you?

One common area of difficulty that is highlighted by candidates is managing time effectively in order to meet deadlines and complete their work (see additional guidance on time management in Chapter 10). A simple but effective way to manage your time is to devise a 'to-do list', (see figure 1.7) highlighting what you need to do and when you need to do it by.

When devising your plan, do not make your plan too long or set unrealistic target dates. All your objectives should be 'SMART':

- **S**pecific – what do you need to do, and why do you need to do it?
- **M**easurable – how will you know when you have achieved it?
- **A**chievable – do you have the skills/knowledge to complete it?
- **R**ealistic – can you really complete it, given the way you usually manage your time and workload? Do not set yourself up to fail by being over-optimistic.
- **T**ime-bound – how long will each task take? When does it need to be completed by?

SMART planning helps when devising a 'to-do list', writing you reflective accounts or getting ready for assessment.

Getting started writing

Planning your written work

One of the things people often find most difficult when studying is actually setting pen to paper and beginning to write. You may have lots of different ideas swimming around in your head or simply not know how to begin. In these situations, and even where you think you have a clear idea of what you need to cover in your written work, it is helpful to begin with a plan. This will often save you time in the long run and will help you to produce a better organised piece of work. Two different methods of planning are an outline or a mind map.

To produce an **outline** begin with a blank sheet of paper and simply write down all of your ideas and thoughts about the things you might need to include as they come into your mind. Do this in note form, and set yourself a time limit, for example 15 minutes, to do so. When you have got all your points on paper, begin to organise them into groups, and consider their importance and relevance. Begin to identify subheadings or sections which would be appropriate for your work, or use the subheadings in one of the models of reflection outlined earlier, and place your points into relevant sections. This will help you to begin to form the structure of your work.

Mind mapping is another method of planning, which also begins by focusing on just getting your ideas on paper. Start by putting the topic of your mind map in a circle in the middle of the page, then draw lines to branch out from it with your main ideas or themes. Smaller 'branches' will help to subdivide the main ideas, and show connections. You may find it helpful to use different colours or illustrations as part of your mind map.

An example of a mind map is provided below. Choose 'Organising my NVQ/SVQ study' or another topic and practice drawing your own mind map.

Important tips for written work

- Use everyday English whenever possible. Imagine you are explaining your point or thoughts to a colleague or manager.
- Do not to write sentences which are too long and become difficult to follow. Try to keep your sentence length down to an average of 15 to 20 words, and try to keep to one main idea in a sentence.
- Explain any abbreviations and technical or legal terms you use. You may want to consider inserting a glossary (a short list explaining the terms) at the end of your work, or in the overall NVQ/SVQ portfolio.
- Divide your work into paragraphs. Each paragraph should relate to one main area or point of discussion. When you begin a new paragraph it indicates that you are moving on to your next point or area of discussion. Your points (and paragraphs) should develop logically and help your work to 'flow'.
- It is often helpful to read what you have written out loud, either to yourself or a colleague. This helps you to identify poor sentence construction and grammar. Also, if you find that your work is sounding like Shakespeare or a romantic novelist, then it is very likely that your language is too complex and flowery for the purposes of your NVQ/SVQ!

Proofreading

Once you have completed your outline or mind map you may find it helpful to leave it for a day or two, and then come back to it and see if there any points you had forgotten, or that you wish to delete.

You should certainly leave a finished piece of written work for a few days wherever possible, and then proofread it thoroughly before submitting it for assessment. Proofread your work when you have

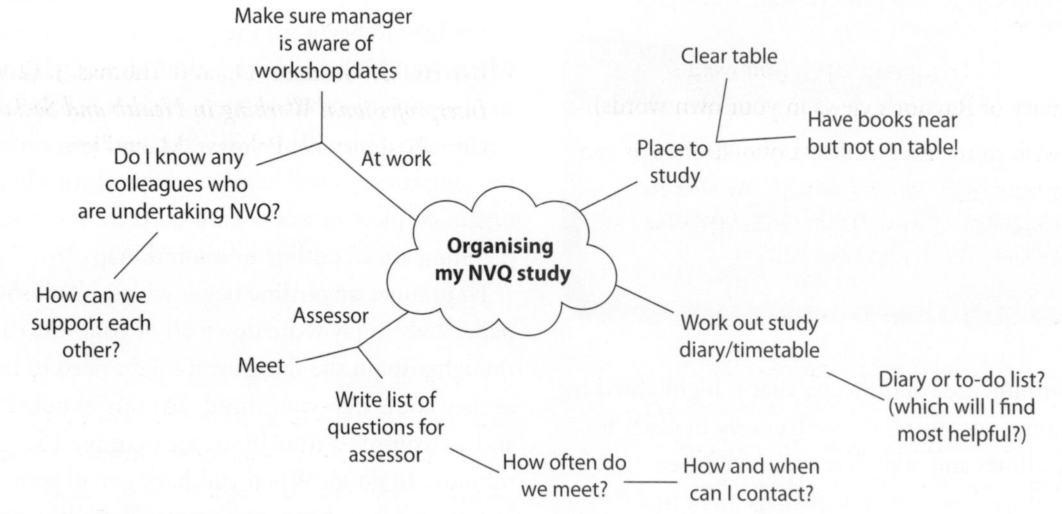

Figure 1.8 Example of a mind map

enough time to do so thoroughly and at a time when you will be able to concentrate fully on the task. Make a conscious effort to read each word and sentence slowly.

Referencing literature in your work

When writing at a higher level, it is important to include appropriate references, to evidence the knowledge which underpins your work. This is also helpful when referencing your evidence to the unit knowledge statements. For example, if you are writing about a piece of legislation or a theory that informs your practice, you need to identify the source of this information. If you do not identify the source of your knowledge, this will be considered as plagiarism.

Including references in your work shows the reader where you obtained your information. It is important that your references are accurate and specific, to allow readers to locate the sources of your information easily, should they wish to do so.

The Harvard referencing system is one of the simplest methods of referencing your work. This system avoids the need to use footnotes or the term 'ibid'.

Referencing in the main body of your work

To reference in the main body of your work, you use the author's last name followed by date of publication:

> Raynor (2001) argues that ... (followed by a summary of Raynor's views in your own words).

If you wish to use a short direct quotation, this can be integrated into the text using quotation marks. If a direct quotation is used, a page number should be included:

> Raynor (2001:82) states that 'Service users ...'

Try to avoid using long quotations, but if you consider one essential, start on a new line and indent the quotation on the page. In this situation, quotation marks are not necessary.

> We should be aware that stress is not simply a matter of occupational stress, but also applies to individuals, families, communities and so on, in any aspects of their life, rather than just the workplace (Thompson, 2002:13).

If there are two authors, cite both: for example, (Brown and Lambert, 2006). If there are more than two authors, use the term 'et al.' to indicate that there are several authors: for example, (Mills et al., 2004).

To reference a book read about in another book – for example, Sapey (2001), which has been read about in Thompson (2001) – use this format:

> ... (Sapey; 2001, cited in Thompson, 2001)

Bibliography or references?

All sources cited in the main text should be listed under a 'References' section at the end of your piece of work. Here you will normally only list items you have specifically mentioned in the main text. If you also list books that have informed your thinking more generally, but which you have not mentioned in the main text, this is a separate 'Bibliography'. The items under both References and Bibliography should be listed in alphabetical order. Note: do not use the shortened form 'et al.' at this point, but give details of all the authors involved.

For a single author, use the following format:

> Hugman, R. (1991) *Power in the Caring Professions*, London: Macmillan.

For joint authors, use:

> Barrett, G., Sellman, D., and Thomas, J. (2005) *Interprofessional Working in Health and Social Care*, Basingstoke: Palgrave Macmillan.

For an edited book, use:

> Thompson, N. (ed) (2001) *Loss and Grief: A Guide for Human Services Practitioners*, Basingstoke: Palgrave.

For a book read about in another book – for example, Sapey (2001), which has been read about in Thompson (2001) – list both authors separately under 'Sapey' and 'Thompson'.

Chapter in an edited book

Normally, edited books should not be included in the bibliography under the editors' names, but be included under the names of the cited authors who wrote the chapters:

Sapey, B. (2001) 'Disability'. In Thompson, N. (ed) *Loss and Grief: A Guide for Human Services Practitioners*, Basingstoke: Palgrave.

Referencing from the Internet

Follow the same format to reference articles from the Internet, but instead of place and name of publisher, you should include the Internet address where you accessed the article and the date on which you accessed it:

Singh, B. (2005) *Making change happen for black and minority ethnic disabled people*, Joseph Rowntree Foundation www.jrf.org.uk/knowledge/findings/socialcare/0495.asp (accessed 10.04.2006).

References

Banks, S. (1995) *Ethics and Values in Social Work*, Basingstoke: Palgrave Macmillan

Beaty, L. (1997) *Developing your teaching through reflective practice*, Birmingham: SEDA

Beverly, A. and Worsley A. (2007) *Learning and Teaching in Social Work Practice*, Basingstoke: Palgrave Macmillan

Boud, D., Keogh, R. and Walker, D. (eds) (1985) *Reflection: Turning Experience into Learning*, London: Kogan Page

Cottrell, S. (2003) *The Study Skills Handbook* (2nd edition), Basingstoke: Palgrave

Gibbs, G. (1988) *Learning By Doing: A Guide to Teaching and Learning Methods*, Oxford: Further Education Unit, Oxford Polytechnic

Johns, C. (1994) 'Guided reflection'. In A. Palmer, S. Burns and C. Bulman (eds) (1994) *Reflective Practice in Nursing: The growth of the professional practitioner*, Oxford: Blackwell Science

Kolb, D. (1984) *Experiential Learning: Experience as a Source of Learning and Development*, New Jersey: Prentice Hall

Krathwohl, D., Bloom, B. and Masia, B. (1964) *Taxonomy of Educational Objectives: The Classification of Educational Goals Handbook II: Affective Domain*, New York: David McKay

McGill, I. and Brockbank, A. (1998) *Facilitating reflective learning in higher education*, Buckingham: Open University

Moon, J. (1999) *Learning Journals: A Handbook for Academics, Students and Professional Development*, London: Kogan Page

Moon, J. (2001) 'The development of assessment criteria for a journal for PGCE students' (unpublished) University of Exeter. In Watton, P., Collings, J. and Moon, J. (2001) *Reflective Writing: Guidance notes for students* www.ex.ac.uk/employability/students/reflective.rtf (accessed 26.8.08.)

Platt, D. (2003) 'Surviving and Thriving in a Changing World.' Speech made at NATOPPS conference 28.4.03 www.dh.gov.uk/en/News/Speeches/Speecheslist/DH_4031709 (accessed 12.9.08.)

Platzer, H., Blake, D. and Snelling, J (1997) 'A Review of Research into the Use of Groups and Discussion to Promote Reflective Practice in Nursing Research'. In *Post Compulsory Education*, 2(2):193–204

Winter, R., Buck, A. and Sobiechowska, P. (1999) *Professional Experience and the Investigative Imagination*, London: Routledge

CHAPTER 2

Leadership and management

'The quality of leadership and management are the key factors driving improvement. What works best is the business-like approach underpinned by the values of social care and an understanding of why it matters to communities.'

(Audit Commission, 2004)

'The issues highlighted by the major enquiries emphasise the crucial role of managers in the care of vulnerable individuals and the delivery of quality services, in conjunction with partners, to meet the needs of these individuals.'

(Watson, 2007:324)

Introduction

This chapter introduces some initial aspects of leadership and management, and begins to consider leadership and management in the care sector. Core elements of leadership and management in the care sector are then considered in more detail in the following chapters. This chapter covers:

- definitions of leadership and management
- theories of leadership and management
- emotional intelligence
- leadership and management in the care sector.

Definitions of leadership and management

The terms 'leadership' and 'management' are often used interchangeably. However, considering definitions of both terms in more detail will help you to highlight the differences between them. Before considering definitions of leadership and management, complete the activity and viewpoint task on page 14.

ACTIVITY

Leadership or management?

The table below lists a number of practice areas. Identify which areas you think are management skills, which are leadership skills, and any areas that are both.

Practice area	Leadership	Management	Both
Ensuring policies and procedures are in place			
Empowering staff to make decisions			
Empowering service users to make choices about their care			
Ensuring all health and safety checks are completed			
Carrying out regular supervision with staff in line with National Minimum Standards			
Ensuring the setting is cost-effective			
Ensuring the setting delivers high-quality services that meet individuals needs			
Facilitating the professional development of staff			

- From your responses, can you identify any differences between leadership and management?
- Were there any areas of practice that you identified as both leadership and management? If yes, why?
- What do you think influenced and informed your choices?

VIEWPOINT

Think of an individual you know, or have known, who you would identify as a good leader. Then think of an individual you know, or have known, who you would identify as a good manager.

- Identify the skills that made them effective as a leader and as a manager.
- Did the manager have good leadership skills, and the leader have good management skills?
- How did their practice influence the way you have developed as a leader and manager?

Many definitions of management focus on activities and what managers 'do', whereas definitions of leadership focus primarily on personal attributes and an ability to inspire and motivate. This can be seen in the definitions below.

Leadership

The *Oxford English Dictionary* defines leadership as 'to guide or to show the way.' Armstrong and Stephens (2005) provide a fuller definition, describing leadership as 'inspiring individuals to give their best to achieve a desired result, gaining commitment and motivating them to achieve defined goals'.

Management

The *Oxford English Dictionary* defines management as 'the professional administration of business concern'. Armstrong and Stephens (2005) identify the role of managers as ensuring 'their organisational department or function operates effectively and are accountable for attaining the required results'.

Stewart (1997) summarises the differences between leadership and management simply: 'Management is essentially about people with responsibility for the work of others and what they actually do operationally, whereas leadership is concerned with the ability to influence others towards a goal.'

REFLECT

- Do you agree with Stewart's distinction between leadership and management?
- Do you think the definition of management fully reflects your role? If not, what is missing?

The Chartered Institute of Personnel and Development (CIPD) (2008a) recognises the connection between the two areas, identifying leadership as an aspect of management, but also reminding us that people who are not managers 'may also function as leaders, influencing others (even if in an informal manner) by their personalities and behaviours'. Denise Platt (2003) reinforces this point with reference to the social care sector:

> 'There are all sorts of leaders in social care, in different parts of our organisations. Our frontline social work staff and frontline care workers provide professional and peer leadership. Our first-line team and group managers provide not only good management but also leadership, which helps people to understand the task and to feel enthusiasm for it.
>
> 'There are leaders who do not have line management responsibilities, many senior practitioners or senior care workers whose role is to generate vision about the best approaches, and to enthuse, and motivate people working in the organisation to do their job to the best of their potential.'

(Speech by Denise Platt, Chief Inspector of the Social Services Inspectorate, 28 April 2003)

As part of an NVQ/SVQ Management workshop, a group of residential managers from a range of settings identified effective leadership and management skills as follows:

Leadership skills

- Inspire and motivate staff
- Good listening skills
- Good communication skills
- Model good practice
- Being creative and innovative
- Being adaptable
- Allowing staff to learn from their mistakes
- Share good practice
- Develop the practice and knowledge of others
- Work well with a wide range of people

Management skills

- Know and understand policies and procedures
- Make sure policies and procedures are carried out
- Able to use range of communication methods
- Can think ahead and plan
- Ensure standards are met
- Have up-to-date knowledge of relevant legislation and standards within the sector.

While the residential managers did identify differences between leadership and management skills, they all felt that to be effective in their role they needed to develop both aspects of their practice and knowledge, and that some skills apply to both leadership and management. This point is reinforced by Skills for Care (2006), which identifies the distinct and overlapping areas in the diagram below.

Leadership
Inspiration
Transformation
Direction
Trust
Empowerment
Creativity
Innovation
Motivation

Common
Communication
Development
Decision making
Integrity
Role model
Negotiation
Knowledge
Professional competence
Setting standards
Flexibility & focus

Management
Delegation
Performance
Planning
Accountability
Finance
Teamwork & team building
Monitoring & evaluating
Formal supervision
Control

Figure 2.1 Common areas of leadership and management

Theories of leadership and management

Some of the main theories of leadership and management are outlined below.

Theories of leadership

Trait theory

This theory suggests that people are born with a range of traits (qualities or attributes), and that some traits are particularly suited to leadership. Individuals who make good leaders have these particular traits. These include:

(Adapted from Stogdill, 1974; Gardner, 1989)

Behavioural theories

Behavioural theories of leadership are based on the belief that great leaders are made, not born – so those who hold a behavioural view would dispute trait theories of leadership. Behavioural theorists argue that it is possible to identify specific behaviours and actions associated with successful leadership, and that people can learn to become leaders through teaching and observation.

Participative theories

Participative leadership theories suggest that the ideal leadership style is one that takes the input of others into account. These leaders encourage participation and contributions from group members, and help group members feel more involved and therefore committed to the decision-making process. Group members will also collaborate rather than compete, as they are working on jointly agreed goals.

A number of individuals have used research studies to identify a range of different leadership styles (Lewin, 1939; Likert, 1961). Some of these styles, including participative, are shown in Figure 2.3.

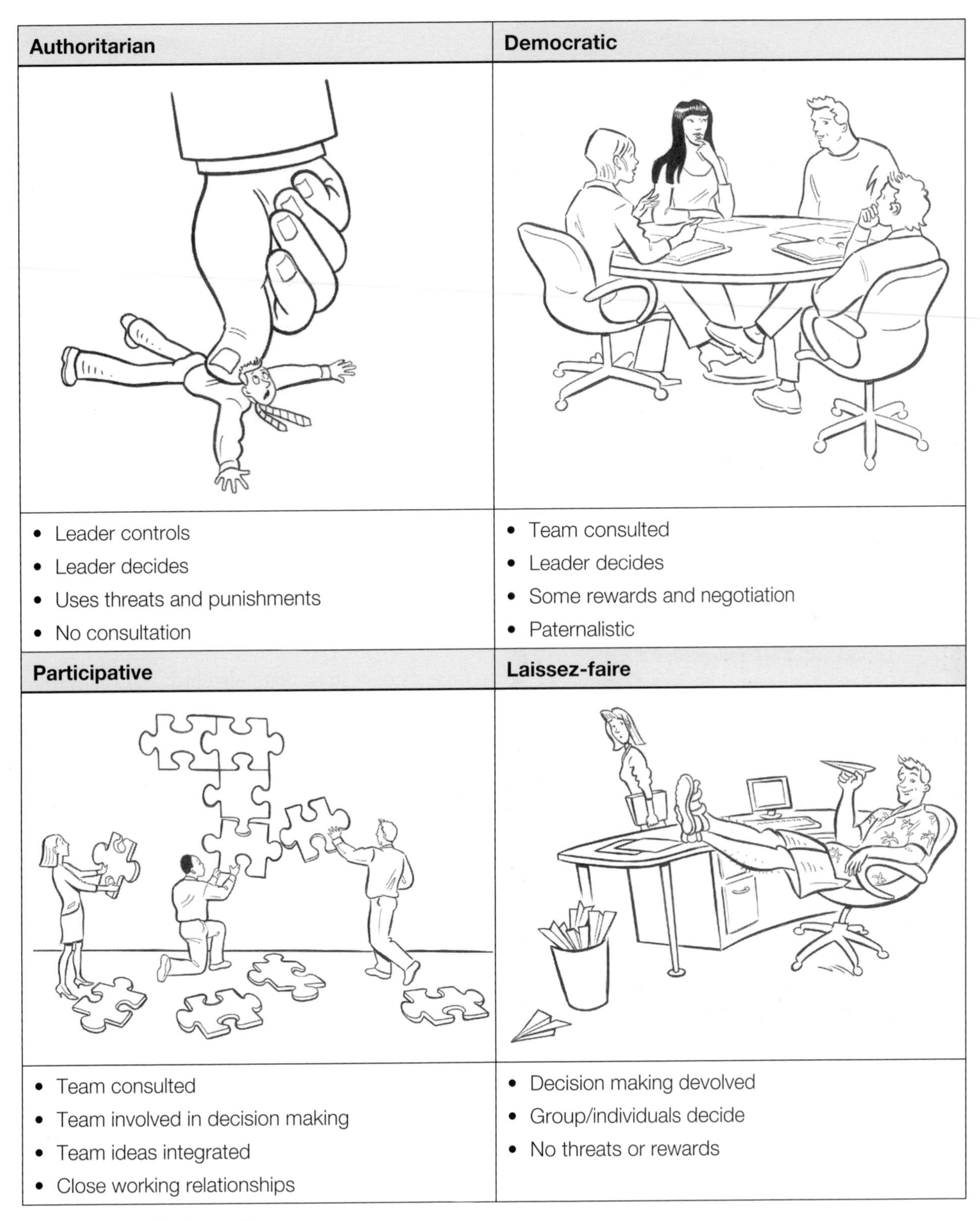

Figure 2.3 Leadership styles

CASE STUDY

Leadership styles

Consider the following scenarios and identify which leadership style is being used.

At a team meeting, ***Angela****, the manager, allocates half an hour for the group to discuss ideas for the home's summer fete. Individuals come up with a range of ideas. Angela thanks everyone for their contributions and says she will confirm the plans for the fete at the next team meeting.*

A few team members are discussing the shift rota and trying to arrange some swaps. ***Bernie****, the manager, overhears the conversation and says all requests must come through him for his decision, and he reminds the team members that his decision will be final.*

The residential home you work in has recently been inspected and a few areas for improvement have been identified. ***Carmen****, the manager, has placed a suggestion box in the reception area so that staff, service users and family members can contribute their ideas of how to improve the areas identified. Carmen has also had three 'surgeries' on evenings and weekends, to give people the opportunity to speak to her directly about their ideas. Next week, you will be involved in the staff and residents' meeting where the ideas will be discussed and the best ones selected.*

It is the end of October and, at a team meeting, several people ask about leave arrangements over the Christmas and New Year period. ***Dougie****, the manager, tells everyone to sort it out among themselves and let him know the shift rota once they have agreed.*

REFLECT

Which leadership style do you use?

Do you use different styles at different times or in different situations? If so, when and why?

Which style do you prefer your manager to use?

Contingency theories

Contingency theories suggest that different situations lend themselves to (or are 'contingent on') a different leadership style, and no particular style is best in all situations. Success depends upon a number of variables, including the leadership style, qualities of the 'followers' (team), and aspects, or context, of the situation.

Transformational theories (also known as relationship theories)

These theories stress that individuals and teams will follow a person who inspires them. Here, leaders motivate and inspire people by helping group members see the importance of the task. Transformational leaders focus on the performance of group members, but also want each individual to fulfil their potential. Some of this thinking links to the concept of 'learning organisations', which is discussed further in Chapter 4 (see page 46).

Theories of management

Scientific management theory

Frederick Taylor developed scientific management theory at the beginning of the 20th century. This theory advocated the specification and measurement of all organisational tasks, which were standardised as much as possible. Although some research showed that this approach appeared to work well for organisations with assembly lines and other routinised activities, it would not be an appropriate approach in residential care settings. Indeed, a routinised approach is identified as one of the elements that can lead to institutional abuse (see Chapter 7, page 122).

Administrative management theory

This theoretical approach focuses on the functions of the manager. Henri Fayol (1916) identified five core elements of management, as shown here.

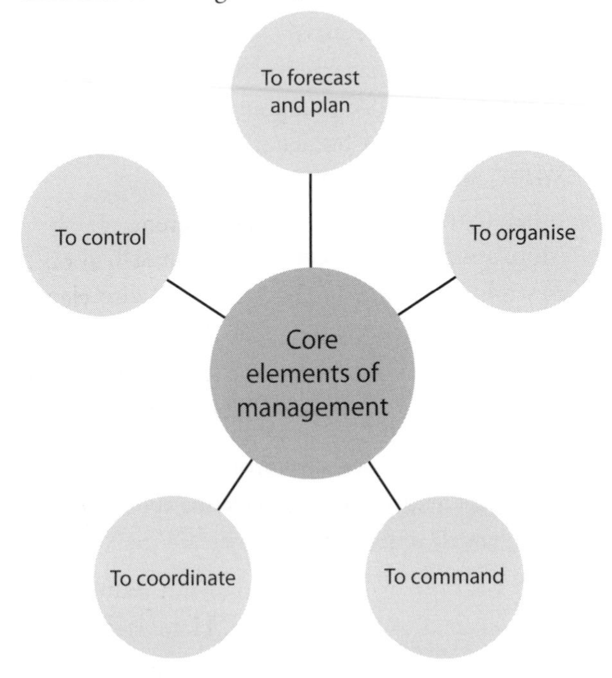

Figure 2.5 Core elements of management

Bureaucratic management theory

Weber defined the key elements of a bureaucracy as including:

- a well-defined hierarchy with a clear chain of command, where higher positions hold the authority to control the lower positions
- the division of labour and specialisation of skills – each employee has the expertise and authority to complete a specific task
- a comprehensive set of rules and regulations
- relationships between managers and employees that are distant and impersonal.

Human relations management theories

These theories focus on the human and behavioural aspects of organisations, which are seen as helping particularly to understand areas such as motivation of staff. Motivation of staff is considered further in Chapter 5 (see page 86).

Systems theory

This theory is more modern (dating from the 1940s onwards). It sees an organisation as being made up of different sub-systems, and recognises the different dynamics of individual organisations, and the impact of internal and external changes on an organisation. Proponents of this theory argue that there is not just one way to manage an organisation, but that management is the process of understanding these dynamics and maintaining effective 'relationships' between the sub-systems of the organisation.

Contingency theory

Like systems theory, contingency theory does not specify a particular management style or strategy, because different situations and issues will require different approaches and solutions. This theory emphasises that managers must be flexible and able to adapt to new situations.

VIEWPOINT

Which, if any, of these management theories seem most relevant to your area of work and why?

Emotional Intelligence

Goleman (1999) argued that emotional intelligence is a crucial component of leadership. Both the Department of Health and Scottish Executive support this view.

Emotional intelligence is 'the capacity for recognising our own feelings and those of others for motivating ourselves, for managing emotions well in ourselves as well as others' (Goleman, 1999:137). Goleman identified five elements of emotional intelligence (see Figure 2.6).

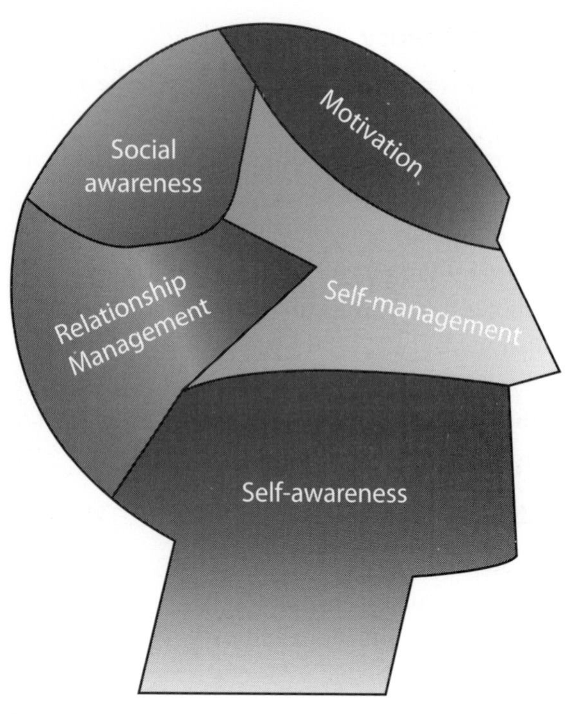

Figure 2.6 Domains of emotional intelligence

- Self-awareness – refers to knowing one's emotions and includes emotional self-awareness, recognising a feeling as it happens as well as self-confidence
- Self-management – refers to the management of one's emotions, and includes handling feelings appropriately, understanding what is behind the feelings, being adaptable and optimistic
- Social awareness – including empathy (sensitivity to others' feelings and concerns, and appreciation of the difference in how people feel) and organisational awareness
- Relationship management – refers to the ability to handle relationships and includes social skills, social competence, influencing and developing others, managing conflict and promoting collaborative working
- Motivation – refers to motivating oneself, controlling feelings and emotions and directing them into achievement of goals

There are critics of Goleman and the concept of emotional intelligence, many arguing that there is a lack of rigorous research supporting it, and that Goleman has made exaggerated claims about the extent to which individuals can increase their emotional intelligence.

CASE STUDY

Using emotional intelligence

Elaine is a care assistant. She has recently gone through a difficult divorce and is the sole carer for her two children, aged 8 and 11. The school holidays are coming up and there is nobody to assist with her childcare. Elaine has come to her manager, Tom, to discuss this. She is rather aggressive and demanding.

Tom is sympathetic to her problem. He recognises that Elaine's aggression is a result of her anxiety about the school holidays. He quickly reassures Elaine and tells her he is sure something can be arranged. Tom agrees to look at the options and talk to her again by the end of the following day.

The next day, Tom meets with Elaine and makes a number of suggestions. He suggests that she could take five weeks unpaid leave during the summer holidays, but spread this across the whole year so that she will still receive a salary each month. He also offers the option of reducing the hours of her working day during term time, if it would help Elaine to take her children to and from school more easily. He tells Elaine to think about these options and come back to him when she has made a decision.

- *In what respects is Tom an emotionally intelligent leader?*
- *What might be the outcome of his suggestions for the organisation and Elaine?*
- *What would you have done?*

(Adapted from Mason, 2003)

CIPD (2007) holds the view that the concept of emotional intelligence is useful but that anyone considering undertaking training in this area, or introducing programmes into the workplace should:

- be aware that there are different models of emotional intelligence (e.g. other than Goleman's model)
- understand that emotional intelligence is not a substitute for 'technical' competence
- appreciate that there are various methods of attempting to develop emotional intelligence
- be sceptical of some of the more inflated claims made for certain emotional intelligence products and techniques
- carefully consider the different 'offers' on the market before buying into particular programmes or approaches.

(CIPD, 2007)

Leadership and management in the care sector

The Scottish Executive (2005) identifies that it should be the manager's role to:

- provide positive leadership
- promote and meet service aims and goals
- develop joint working partnerships that are purposeful and effective
- allow staff and service users and carers to develop services that people want
- value people by recognising and actively developing potential
- develop and maintain awareness of learning and development and make sure that time is allowed and processes are in place for listening to staff, service users and carers
- provide an environment that allows time to reflect upon practice
- develop professional skills and the ability to make judgements in complex situations, making sure staff have access to professional consultation, support and advice
- support changes that result from critical evaluation of practice and regular performance appraisals
- take responsibility for the continuing development of staff.

REFLECT

- Are there areas in the list above that you would identify as specific to the care sector? If so, which are they and why?
- Is there anything which you consider is missing from the list? If so, what?

These roles begin to reflect the view that leadership and management in social care (or service provision) organisations differs from that in more 'industrial' organisations (Wilderom, 1991). The differences are seen to arise primarily because of the imperative in social care for direct interaction with service users to set goals and strategies. There, the need for collaborative or joint working and an ability to support changes is also highlighted. There is also a particular emphasis on the empowerment of others (service users and staff). So leadership in social care organisations may be seen as more complex, but management practice in this area is also underpinned by a specific value base. That is not to say that businesses in other sectors do not have a value base, but this must be core within any organisation operating in the care sector. The quotation from the Audit Commission at the beginning of the chapter helps to reinforce this point, as do these.

> 'What is needed are managers with a clear set of values about the role of public services, particularly in relation to addressing the needs of vulnerable people, combined with the ability to "lead from the front".'

(Laming, 2003:6)

> 'Leadership and management practice must integrate social care values. Leaders and managers must have a critical understanding of anti-discriminatory practice and the impact of

exclusion and disadvantage on people's lives. Leaders and managers must be proactive in working with the diversity of staff, of people who use services and of carers, and in promoting social inclusion.'

(Skills for Care, 2006:9)

Skills for Care also remind us that leaders and managers in social care should be able to implement the Codes of Practice, particularly the requirements to:

- protect the rights and promote the interests of people who use services and of carers
- strive to maintain the trust and confidence of people who use services and of carers
- promote the independence of people who use services and protect them as far as possible from danger or harm
- balance the rights of people who use services and of carers with the interests of society
- uphold public trust and confidence in social care services.

They also remind us, as employers, to:

- regularly supervise and effectively manage staff to support good practice and professional development and to address any deficiencies in their performance
- provide training and development opportunities to help staff to do their jobs and to strengthen and develop their skills and knowledge
- provide a safe working environment in which dangerous, discriminatory or exploitative behaviour are known to be unacceptable and are addressed.

(Skills for Care, 2006)

The following chapters of this book are underpinned by, and integrate, these core social care values and codes of practice. These should be reflected in all aspects of your own practice.

NVQ/SVQ

This chapter links to all units as it considers aspects of leadership and management that underpin the criteria in all units. The chapter will also help you to address the knowledge statements within the award.

Work Products you may generate and either include in your portfolio or show your assessor to demonstrate your skills and knowledge are:

- **appraisal and supervision records that refer to your leadership and management skills**
- **records of training and development in leadership and management**
- **learning logs or reflective accounts evidencing the learning from these events.**

References

Armstrong, M. and Stephens, T. (2005) *A Handbook of Leadership and Management*, London: Kogan Page

Audit Commission (2004) *Old Virtues, New Virtues: An overview of the changes in social care services over the seven years of Joint Reviews in England 1996–2003*, Wetherby: Audit Commission Publications

CIPD (2007) *Emotional Intelligence*, London: CIPD

CIPD (2008a) *Leadership: an overview*, London: CIPD

Gardner, J. (1989) *On Leadership*, New York: Free Press

Goleman, D. (1999) *Working with Emotional Intelligence*, London: Bloomsbury

Lord Laming (2003) *The Victoria Climbié Inquiry: Summary Report of an Inquiry*, Cheltenham: HMSO

Lewin, K., Lippit, R. and White, R.K. (1939) 'Patterns of aggressive behavior in experimentally created social climates', *Journal of Social Psychology*, 10: 271–301

Mason, L. (2003) 'Leading Teams in the 21st Century'. In Thomas, A. (2003) *Leading and Inspiring Teams* Oxford: Heinemann Educational

NHS Leadership Centre (2003) *NHS Leadership Qualities Framework* www.nhsleadershipqualities.nhs.uk (03.09.08)

Platt, D. (2003) 'Surviving and Thriving in a Changing World'. Speech made at NATOPPS conference 28.4.03 www.dh.gov.uk/en/News/Speeches/Speecheslist/DH_4031709 (accessed)

Scottish Executive (2005) *National Strategy for the Development of the Social Service Workforce in Scotland: A Plan for Action 2005–2010*

Scragg, T. (2001) *Managing at the Front Line: A handbook for managers in social care agencies*, Brighton: Pavilion

Skills for Care (2006) *Leadership & Management Strategy: A Strategy for the Social Care Workforce*, Leeds: Skills for Care

Stewart, R. (1997) (3rd ed) *The Reality of Management*, Oxford: Butterworth-Heinemann

Stogdill, R.M. (1974) *Handbook of Leadership: A survey of the literature*, New York: Free Press

Watson, Joan E.R. (2007) '"The Times They Are A Changing" – Post Qualifying Training Needs of Social Work Managers', *Social Work Education*, 27(3):318–333

Wilderom, C.P.M. (1991) 'Service management/leadership: different from management/leadership in industrial organisations?', *International Journal of Service Industry Management*, vol 2(1):6–14

Useful reading

Collins, Jim (2005) *Good to Great and the Social Sectors* Colorado: Jim Collins

Skills for Care (2006) *Leadership & Management Strategy: A Strategy for the Social Care Workforce* Leeds: Skills for Care

CHAPTER 3

Working with service users, families and carers

'The totality of people's lives needs to be understood and appreciated; service users are not a burden, so help them make a contribution. Good support is a sign of a good society. We just want to have a normal life and be able to do the things that everybody else does and take for granted.'

(Beresford, 2005:1)

Introduction

A key value underpinning this chapter, and the book overall, is that of person-centred care. The chapter will focus on each of the areas below from this perspective. The chapter covers:

- holistic and person-centred care
- medical and social models of disability
- the legal and policy context of person-centred care
- theories of adult development
- person-centred planning, including care planning and reviews
- managing information.

Holistic and person-centred care

'Holistic' and 'person-centred' care are terms commonly used in the health and care sector. This chapter aims to help you to identify what this type of care 'looks like', and how to achieve it in your organisation and with those you work with.

As indicated in the quotation that opens this chapter, individuals' needs must be considered holistically: that is, considering the 'whole person'. In this way, all aspects of an individual and their needs – physical, psychological, social, spiritual and other needs – would be considered.

Care should also be person-centred, meaning that care 'is user-focused; promotes independence and autonomy rather than control; involves users choosing from reliable, flexible services;

and tends to be offered by those working with a collaborative/team philosophy' (Innes et al., 2006).

Person-centred care aims to see the person as an individual, rather than focusing on a condition or illness, or on abilities they may have lost. The Alzheimer's Society (2008) comments that 'instead of treating the person as a collection of symptoms and behaviours to be controlled, person-centred care takes into account each individual's unique qualities, abilities, interests, preferences and needs'.

Figure 3.1 Person-centred care sees the individual

Medical and social models of disability

The traditional model of disability is often referred to as the 'medical model' of disability. This model views an individual's disability as a medical 'problem', which results in individual not being able to function fully in society because of their disability. The medical model 'implies that the way to overcome barriers to inclusion is to "adjust" the individual disabled person in order to "fit" society, rather than adjust society to accommodate disabled people' (Elder-Woodward, 2005).

The social model of disability (Figure 3.2) was developed by disabled people themselves and argues that it is society itself that prevents individuals from being full and active participants because of the barriers it erects. 'It is not individual limitations, of

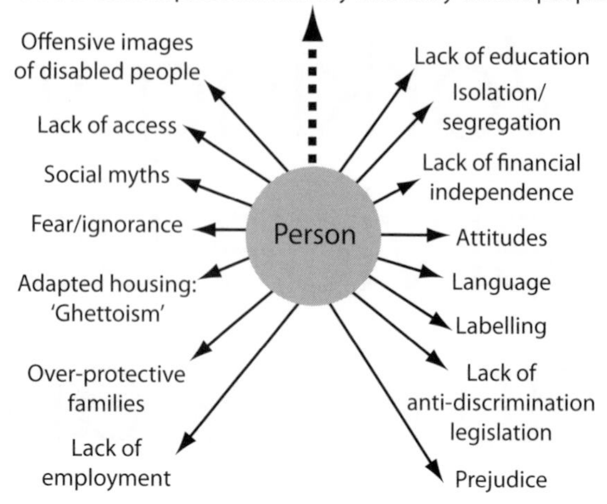

Figure 3.2 Social model of disability

ACTIVITY

Models of disability

Read each of the questions below, and decide whether they are based on the medical or social model of disability.

Question	Medical or social model?
Does your health problem/ disability prevent you from going out as often or as far as you would like?	
Are there any transport or financial problems which prevent you from going out as often or as far as you would like?	
What is it about the local environment that makes it difficult for you to get about?	
Does your health problem/ disability make it difficult for you to travel by bus?	

(Adapted from Demos, 2003, http://jarmin.com/demos/course/awareness/oliver.html)

whatever kind, which are the cause of the problem but society's failure to provide appropriate services and adequately ensure the needs of disabled people are fully taken into account in its social organisation' (Oliver, 1990). This view of disability therefore also sees society's approach to disability as a form of discrimination, which the Disability Discrimination Act 1995 (mentioned in the next section) seeks to address.

The social model of disability is clearly more person-centred than the medical model.

VIEWPOINT

Review any documentation which asks questions of service users (such as assessment documents) and identify any questions which are phrased using the medical model. Is it appropriate that the question(s) is phrased in this way? If not, rephrase the question(s) drawing on the social model.

ACTIVITY

Adaptations to your setting

'When I had to start using a wheelchair, it upset me that I couldn't get out into the garden. It was easily solved by the manager, who arranged for a ramp to be built from the door by my room down to the garden'

(CSCI, 2005:6).

This quote is a real example of a manager applying the social model and adapting a residential home to meet the needs of an individual service user. Identify two examples of circumstances where you have made adaptations within your setting to meet the particular needs of a service user.

- What adaptations did you make and why?
- Did you encounter any difficulties in making adaptations, and if so, how did you overcome them?

Legal and policy context of person-centred care

The service users who contributed to the report *Developing social care: service users' vision for adult support* felt strongly that social care should be based on the social model 'informing policy, practice and procedures' (Beresford, 2005). The Department of Health clearly states that it also subscribes to the social model of disability (DOH, 2008a).

There is a range of current legislation and policy underpinned by the social model, directing and supporting the provision of person-centred care. As the manager, you need to ensure that you are aware of current legislation and policy and how they apply to your setting and work.

Human Rights Act 1998

The Human Rights Act 1998 is derived from the European Convention on Human Rights, which the UK signed in the 1950s, and came into force in October 2000. The 1998 Act incorporates articles from the Convention such as:

- Article 5 – The right to liberty and security
- Article 8 – The right to respect for private and family life
- Article 9 – The right to freedom of thought, conscience and religion.

The Human Rights Act 1998 should be used as a basis for good practice in all social care settings. It underlines that the needs of the individual should be paramount in all areas and that everyone has the right to receive 'quality' care services. 'The dignity, needs and wants of older people must be put at the centre of their care, and Human Rights are the perfect vehicle to ensure this and to deliver quality care services' (Age Concern, 2008).

Equalities Act 2006

The Equalities Act 2006 places a statutory duty on public authorities to ensure they:

- eliminate unlawful discrimination and harassment
- promote equality of opportunity for/between men and women.

The Act makes it unlawful to discriminate on the grounds of a person's sexual orientation, religion and beliefs in the following areas:

- the provision of goods, facilities and services
- the disposal and management of premises
- education
- the exercise of public functions.

The Equalities Act 2006 also established a single Commission for Equality and Human Rights (CEHR), replacing the Equal Opportunities Commission (EOC), the Commission for Racial Equality (CRE) and the Disability Rights Commission (DRC). The CEHR is responsible for promoting understanding of equality and human rights issues and for challenging unlawful discrimination. It has responsibility for promoting understanding of areas of discrimination such as sexual orientation, religion or belief and age. It also has a wider remit to promote human rights and equality generally, even those areas not covered by specific pieces of legislation.

Disability Discrimination Act 1995

The Disability Discrimination Act 1995 aims to stop the discrimination many disabled people face and to give them equal rights and access to goods, facilities and services. The Act says that service providers are not allowed to treat disabled people less favourably because of their disability. Under the Act, 'service providers' are most companies or organisations that offer goods, facilities or services to the public in the UK. It does not matter whether the services are free or paid for, or whether they are provided by the public or private sector.

NHS and Community Care Act 1990

One aspect of the NHS and Community Care Act 1990 is the requirement for local authorities to carry out 'needs-led' assessments and to show that it has taken a person's preferences into account (although not necessarily followed them). Needs-led assessments are person-centred, whereas 'resource-led' assessments are driven by the availability of resources from a particular service or area. For example, an individual may be offered a place at a day centre because that is all that is available, when a 'sitting' service would be more appropriate to a service user and carer's needs.

When working in a context of often-expanding demand and only finite resources, it is important to acknowledge the challenges that exist in promoting and achieving needs-led assessments.

Care Standards Act 2000

The regulations introduced under the Care Standards Act 2000 aim to protect the privacy and dignity of people receiving care in their own homes or in care homes. The Care Standards Act 2000 itself put in place a new independent regulatory system regarding care services. The main body regulating and inspecting care services in England, the Commission for Social Care Inspection (CSCI), identify the first of their four key aims which guide their work as: putting the people who use social care first. The CSCI will be replaced by the Care Quality Commission from 2010.

Community Care (Direct Payments) Act 1996

The Department of Health defines direct payments as 'cash payments made to individuals who have been assessed as needing services, in lieu of social service provisions' (DOH, 2008). Direct payments were first introduced in April 1997 for adults of working age, but have since been extended to other groups, including older people and people with short-term needs (for example, someone who needs help at home while recovering from an operation). Councils must offer and make a direct payment to eligible individuals who are able to provide consent. The law also specifies that direct payments should be discussed as a first option at each assessment and each review.

The government believes that 'by giving individuals money in lieu of social care services, people have greater choice and control over their lives, and are able to make their own decisions about how their care is delivered' (DOH, 2008).

You may also come across the terms 'personal budget' and 'individual budget'. The key difference between personal budgets and individual budgets is that the personal budgets are solely from adult social care funding, while individual budgets bring together a number of funding and support streams in addition to adult social care (for example, local authority-provided social care, independent living fund, disabled facilities grant and integrated community equipment services). An individual budget can cover more than personal social care – for example, it could cover access to work – and can be a cash payment, or arranged services, or a combination of both.

Mental Capacity Act 2005

The Mental Capacity Act was implemented in October 2007, and seeks to empower and protect people who lack the capacity to make their own decisions. There are five core principles underpinning the Act.

- It is assumed that individuals have capacity, unless an assessment has taken place that determines otherwise.
- Individuals must be given appropriate support to make a decision (before a judgement can be made that a person cannot do so).
- Individuals have the right to make an unwise decision: just because an individual makes a decision that appears unwise, it does not mean that they lack the capacity to do so.
- If it is determined that an individual lacks capacity, any decision made on their behalf must be made in their best interests.
- Any actions taken in respect of the individual without capacity should be the least restrictive in terms of their basic rights and freedoms.

The Department of Health has produced some useful training materials on the Mental Capacity Act (2005), which can be accessed from their website.

ACTIVITY

Three Stories

Watch the 'Three Stories' film vignettes on http://www.publicguardian.gov.uk/about/three-stories.htm

- Identify your key learning points from this film.
- Which parts of the video are most relevant to your setting and why?
- How might you use this with staff to increase their understanding of the Act?

REFLECT

Can you identify any other legislation which you apply in your setting and which directs and supports the provision of person-centred care?

The table below (Figure 3.3) outlines a range of the key policies that also relate to person-centred care.

Policy	**Link to person-centred care**
Valuing People (DOH, 2001)	The White Paper *Valuing People* stresses that a 'person-centred approach will be essential to deliver real change in the lives of people with learning disabilities'. As part of this approach, all organisations working with people with learning disabilities have to implement person-centred planning for all service users.
National Service Frameworks and the Single Assessment Process	National Service Frameworks (NSFs) are long-term strategies for improving specific areas of care, and person-centred care is a theme which runs through the NSFs. The Single Assessment Process (SAP) was introduced in the National Service Framework for Older People under Standard 2: Person-Centred Care. The aim of this standard is 'to ensure that older people are treated as individuals and they receive appropriate and timely packages of care which meet their needs as individuals, regardless of health or social boundaries' (DOH, 2001). The SAP aims to ensure that older people's needs are assessed holistically, with professionals working together. This should avoid the range of professionals working with, or providing services for, an individual carrying out their own assessments. The success of SAP is therefore dependent on the different professionals successfully working together (see Chapter 9).
Care Programme Approach (CPA)	The CPA was introduced in 1990, to provide a framework for effective mental health care for people with severe mental health problems. There are similarities between the CPA and SAP – specifically that service users need to be fully involved in the planning process and that professionals need to work together to ensure that the service users' needs are met.
National Minimum Standards	National Minimum Standards constitute the minimum expectations the government sets for care providers in the services they deliver. National Minimum Standards are not legally enforceable but are guidelines for providers, commissioners and users to assist them in judging the quality of a service. Some standards relate to person-centred planning: for example, *The National Minimum Standards Care Homes for Adults* (18 – 65) (DOH, 2003) Standard 6.1 states that 'the registered manager develops and agrees with each service user an individual Plan, which may include treatment and rehabilitation, describing the services and facilities to be provided by the home, and how these services will meet current and changing needs and aspirations and achieve goals'.
Independence, Well-being and Choice (DOH, 2005) *Our Health, Our Care, Our Say* (DOH, 2006)	These Green and White Papers saw the continuation of government policy developments that have since resulted in *Putting People First* (see below). *Independence, Well-Being and Choice* set out the government's vision for adult social care for the 10–15 years following its publication in 2005. *Our Health, Our Care, Our Say* builds on this and also incorporates health services. Both documents are underpinned by person-centred approaches to all aspects of care provision.

Policy	Link to person-centred care
The Personalisation Agenda and *Putting People First*	The government summarises its approach to personalisation as 'the way in which services are tailored to the needs and preferences of citizens. The overall vision is that the state should empower citizens to shape their own lives and the services they receive' (DOH, 2008). *Putting People First* is an agreement between six government departments, the Local Government Association, the Association of Directors of Adult Social Services, the NHS, representatives of independent sector providers, the CSCI and other partners. It sets out the aims and values shared by these groups (based on the Personalisation Agenda) that will underpin their work in developing adult social care. The Department of Health explains personalisation as meaning that 'everyone who receives social care support, regardless of their level of need, in any setting, whether from statutory services, the third and community or private sector or by funding it themselves, will have choice and control over how that support is delivered. It will mean that people are able to live their own lives as they wish, confident that services are of high quality, are safe and promote their own individual requirements for independence, well-being and dignity' (DOH, 2008d).

Figure 3.3 How key policies relate to person-centred care

Dignity in Care

Dignity in Care is a campaign launched by the Department of Health in 2006 which seeks to eliminate the tolerance of indignity in health and social care services. Although this is a 'campaign' rather than a policy document, it is mentioned here because it also supports a person-centred approach to care.

The Department of Health has laid out a 'Dignity Challenge' identifying the national expectations of what constitutes a service that respects dignity. It focuses on ten different aspects of dignity which are the things that people have reported as mattering most to them.

The Dignity Challenge

High-quality care services that respect people's dignity should:

1. have a zero tolerance of all forms of abuse
2. support people with the same respect you would want for yourself or a member of your family
3. treat each person as an individual by offering a personalised service
4. enable people to maintain the maximum possible level of independence, choice and control
5. listen and support people to express their needs and wants
6. respect people's right to privacy
7. ensure people feel able to complain without fear of retribution
8. engage with family members and carers as care partners
9. assist people to maintain confidence and a positive self-esteem
10. act to alleviate people's loneliness and isolation.

REFLECT

- How well do you think your service currently meets the Dignity Challenge?
- In which aspects is your service strongest?
- In which aspects is your service weakest? How could you improve in these areas?

Theories of adult development

This section of the chapter outlines a number of theories related to adult development. Only a selection of theories is presented here, and you may be aware of and draw on other theories in your work. Understanding theories of adult development will help you to consider and possibly understand people's lives, experiences, current situation and needs more fully.

Erikson's stages of development

Erik Erikson (1902–94) developed a theory that divides an individual's life into eight stages that extend from birth to death (unlike many developmental theories that only cover childhood). Erikson believed that stages of a person's development (see Figure 3.4) are linked to their social and cognitive development, rather than purely being led by their physical needs. He was also interested in how the culture and society an individual lives in could influence their development. Erikson believed that basic conflicts are encountered at each stage of development and that, if these are not fully resolved at that stage, they can have a cumulative negative effect in later stages of development.

Stage of development	Basic conflict at this stage	Summary
Infant	*Trust vs. Mistrust*	The infant must form a first loving, trusting relationship with the caregiver, or develop a sense of mistrust
Toddler	*Autonomy vs. Shame and Doubt*	The child works to master their physical environment while striving to maintaining self-esteem
Preschooler	*Initiative vs. Guilt*	The child begins to initiate, not imitate, activities, and develops conscience and sexual identity
School-age child	*Industry vs. Inferiority*	The child tries to develop a sense of self-worth by refining skills
Adolescent	*Identity vs. Role Confusion*	The child tries integrating many roles (e.g. child, sibling, student, worker) into a self-image under role-model and peer pressure
Young adult	*Intimacy vs. Isolation*	The adult learns to make personal commitment to another as spouse, parent or partner
Middle-age adult	*Generativity vs. Stagnation*	The adult seeks satisfaction through productivity in career, family, and civic interests
Older adult	*Integrity vs. Despair*	The adult reviews their life accomplishments, deals with loss and prepares for death

Figure 3.4 Erikson's stages of development

CASE STUDY

Read through the four case studies below and identify which of Erikson's stages of development and conflicts each individual is going through.

Billy *is a young man with learning disabilities. He is very sociable and has a good group of friends, and has had several girlfriends in the past, but only for short periods. He tells you that he is tired of dating different girls and wants to have one special 'forever' girlfriend.*

Florence *can't wait for her grandchildren's next visit. She loves to spend time telling them stories about their mother when she was their age. She talks about the home they lived in and holidays they enjoyed together as a family. Sometimes, she says she could have been a famous singer, but she acknowledges that, if she had kept trying to pursue that dream, she 'wouldn't have had time to have your mum!'*

Stanley *is an 82-year-old widower who lives in a sheltered housing scheme. Stanley is in relatively good health, but is frail. He rarely goes out and does not take part in activities offered, staying in his flat instead. Other residents in the scheme say 'it is better that way' because all he does is moan about how difficult his life has been, and how he has 'had things worse' than most people.*

Mia *is 33 and went through a residential drug rehabilitation programme for the second time three months ago. She is currently attending the after-care day service programme offered by the unit. She is making good progress in getting things back on track. She has just booked an appointment with a careers adviser to help her to decide what to do 'with the rest of her life'.*

Maslow's Hierarchy of Needs

Abraham Maslow (1908–70) was an American psychologist who developed a theory of a Hierarchy of Needs. This suggested that an individual's needs must first be met at a basic level, and then must be satisfied at each level before moving onto the next. The diagram below (Figure 3.5) outlines the individuals needs at each level.

Although individuals may reach a point where most of their important needs are met, Maslow believed that only a minority of people reach the top of the hierarchy. Also at different times, if individuals are having difficulties in their lives, they may regress to lower levels of the hierarchy, prioritising their more basic needs. For example, if an individual were made redundant, they would be more likely to be concerned with securing another job to enable them to keep their home and buy food, than with the particular status or recognition their next job might bring.

Maslow also identified that problems or difficult circumstances at one point in a person's life might cause them to 'fixate' on a particular set of needs, possibly affecting their future happiness. For example, a person who suffered extreme deprivation and lack of security in early childhood may fixate on physiological and safety needs. These may remain significant to the individual so that, even if the individual later has a successful, well-paid and secure job, they may still fixate over keeping enough food in their fridge.

REFLECT

- Identify at least two ways in which your setting supports service users at each stage of Maslow's hierarchy.
- What particular difficulties might your setting face in helping service users to achieve the higher levels of need? Why – and what could you do to overcome this?

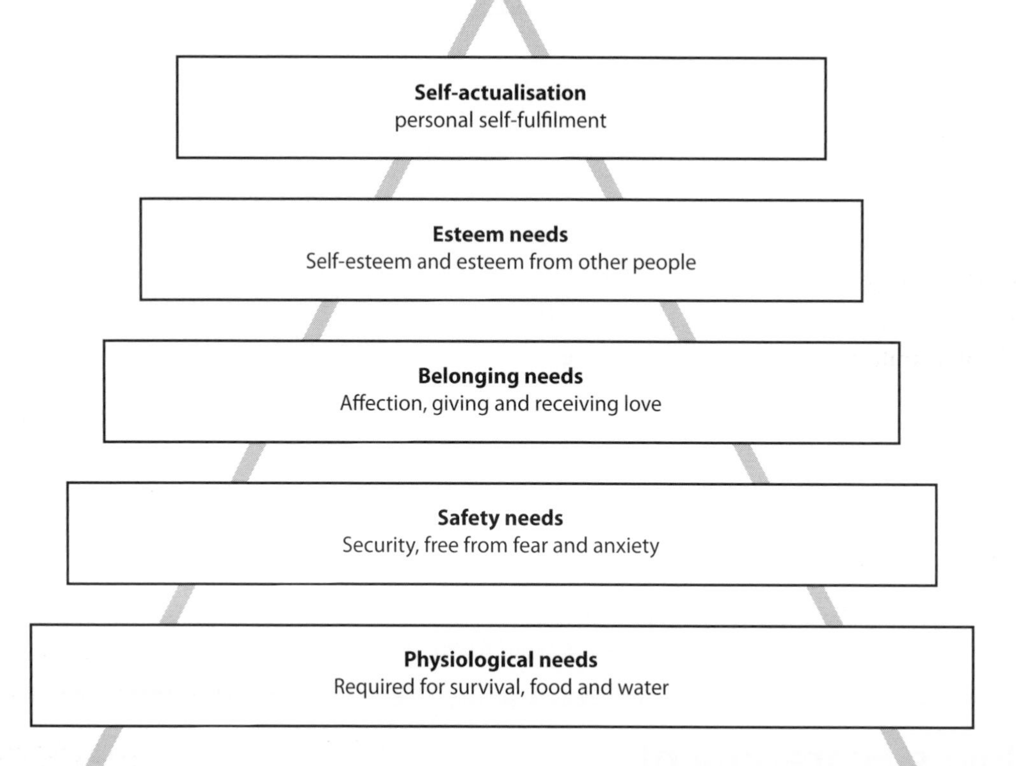

Figure 3.5 Maslow's Hierarchy of Needs

Kübler-Ross' stages of grief

Elisabeth Kübler-Ross (1926–2004) developed a theory that identified five stages of grief individuals move through when they suffer a loss. She used this understanding of the grief process to develop approaches in the support given to those who have encountered trauma and grief associated with loss.

The five stages

1. **Denial** – At this stage, the individual refuses to accept their loss. The loss may feel 'unreal' and that 'this is not happening to me'.
2. **Anger** – The individual may become angry at those around them, because of the 'unfairness' of their loss. This anger can be overwhelming and consuming. The individual may be asking questions such as 'Why did this happen to me?'
3. **Bargaining** – At this stage, the individual is seeking a way out of their situation, often bargaining with their 'god' or a 'higher power' to reverse the loss in return for the individual making changes, for example, in their behaviour or lifestyle.
4. **Depression** – Individuals are just beginning to realise that their situation is irrevocable and they really must continue to live without the presence of the person or aspect of their lives that has been lost. Individuals may also blame themselves for having caused or contributed to their loss in some way, whether or not this is justified. During this phase, the individual may feel sad, or anxious and even some regret or guilt.
5. **Acceptance** – Individuals will move to the acceptance stage after they have become comfortable with the fact that the person who has died, or aspect of life which has been lost, will not be returning to them. This acceptance may also

result in the individual deciding to make significant life changes, such as moving house.

This model could be related to the variety of losses a service user and their carers may encounter. These could include the death of a relative or a service user within the setting, but may relate to losses other than bereavement. For example, it could concern the change of a key worker or residential setting the service user is used to, or other life-changing experiences, such as declining health.

Holmes (2004:179) suggests 'there is strong evidence of the relationship between acute loss and increased vulnerability to psychiatric and physical disorder'. This reinforces how vulnerable individuals can be when grieving, and the importance of offering appropriate levels of support.

VIEWPOINT

Think of a situation where you experienced a loss related to your work: for example, a close colleague leaving the organisation or the loss of a role you valued.

- How did you cope with the loss?
- Did the situation affect the way you were able to work with and support service users, staff and colleagues? How?
- How did the structures and systems in your organisation support you during this situation?
- Where did you, or could you have, sought additional support from?

Coping mechanisms

Supporting those who have experienced loss can be stressful and traumatic. As the manager of a setting, you will get to know those who reside in your setting and may be affected as much as others by any loss, but you are also the one others look to for support and guidance during difficult periods. Everyone needs to have coping mechanisms in place to help them during times of stress and loss. Use the

CASE STUDY

Frank

Frank is a widower who moved into your care setting last year. His wife had died six months earlier and Frank could no longer cope at home. He has early stages of dementia. Puddles, the cat which belonged to Frank's wife, also moved into the home with him. Puddles provides great comfort to Frank and has become a favourite with the other residents. You arrive at work one morning to discover that Puddles has been run over and has died.

- *How will you support Frank through this loss?*
- *How will you support other residents?*

Simone

Simone has learning difficulties and lives in a semi-independent unit which is part of the service you manage. Simone's younger sister Anya visits her frequently and they are very close. Anya is sitting her 'A' levels in a few months and, assuming she is successful in her grades, will be going away to university in September. This means Simone will see Anya much less often.

- *How will you work with Simone and the staff team to support her in the periods before and after Anya goes to university?*
- *How might you support Frank and Simone at each of the stages of grief identified by Kübler-Ross?*
- *What strategies would you use at each stage and why?*
- *What other factors might you need to keep in mind in each situation?*

viewpoint on page 35 to identify your own methods of coping with loss, and sources of support which you could access.

Socio-cultural aspects of adult development

Socio-cultural aspects such as class, gender, ethnicity, religion and sexual orientation influence adult development and identity (Chavez and Guido-DiBrito, 1999; Cross, 1995; Kroger, 1997; Merriam and Caffarella, 1999). It is important to note that many of the dominant adult development theories have been devised from a white, middle-class, American and European perspective. These theories have therefore been developed within particular value systems, and in relation to a limited range of cultures. This raises questions about the validity of the dominant adult development theories in relation to a significant proportion of service users and carers. It also highlights possible issues of cultural bias and misunderstanding, which would impact on person-centred care.

REFLECT

- Write down five words or phrases that identify your culture and identity.
- Consider how your culture and identity might have been shaped if you had been brought up in a different country.
- What is the dominant culture in your work setting? Write down words and phrases to describe this.
- How does this impact on you, service users and staff?
- How do you gather information from service users regarding their culture and identity?
- Having completed this reflection, would you view or apply the theories of Erikson, Maslow and Kübler-Ross differently? If so, why and how?

As stated at the beginning of this section, only a limited selection of theories of adult development has been outlined here. Consideration of the socio-cultural aspects of development highlights that it is not possible to select one theory and apply it wholesale to every individual, or every situation. Theories can only assist in developing our understanding of individuals; each person's life, circumstances, experiences and perceptions of these areas will determine the individual they are, their needs, wishes and feelings. It is only with this in mind that our practice can be 'person-centred'.

Person-centred planning (PCP)

'Person-centred planning is a process of life planning for individuals, based around the principles of inclusion and the social model of disability' (The Circles Network, 2008).

A key element of *Valuing people* (DOH, 2001), person-centred planning is a term used in relation to planning with services users who have a learning disability. This section will refer primarily to documents, and use examples, related to this service user group. However, the Single Assessment Process and Care Programme Approach mentioned earlier are also based on the principles of person-centred planning, although they are also concerned with the allocation of resources and eligibility criteria.

PCP means that the service user is fully involved in all aspects of their care and reinforces that service providers need to ensure they listen to what people really want.

PCP has five key features.

1. The person is at the centre.
2. Family and friends are full partners in planning.
3. A plan identifies what is important to the person (now or for the future). It identifies their strengths and what support they need
4. A plan helps the person to be part of their community. It is not just about services and shows what is possible, not just what is on offer.

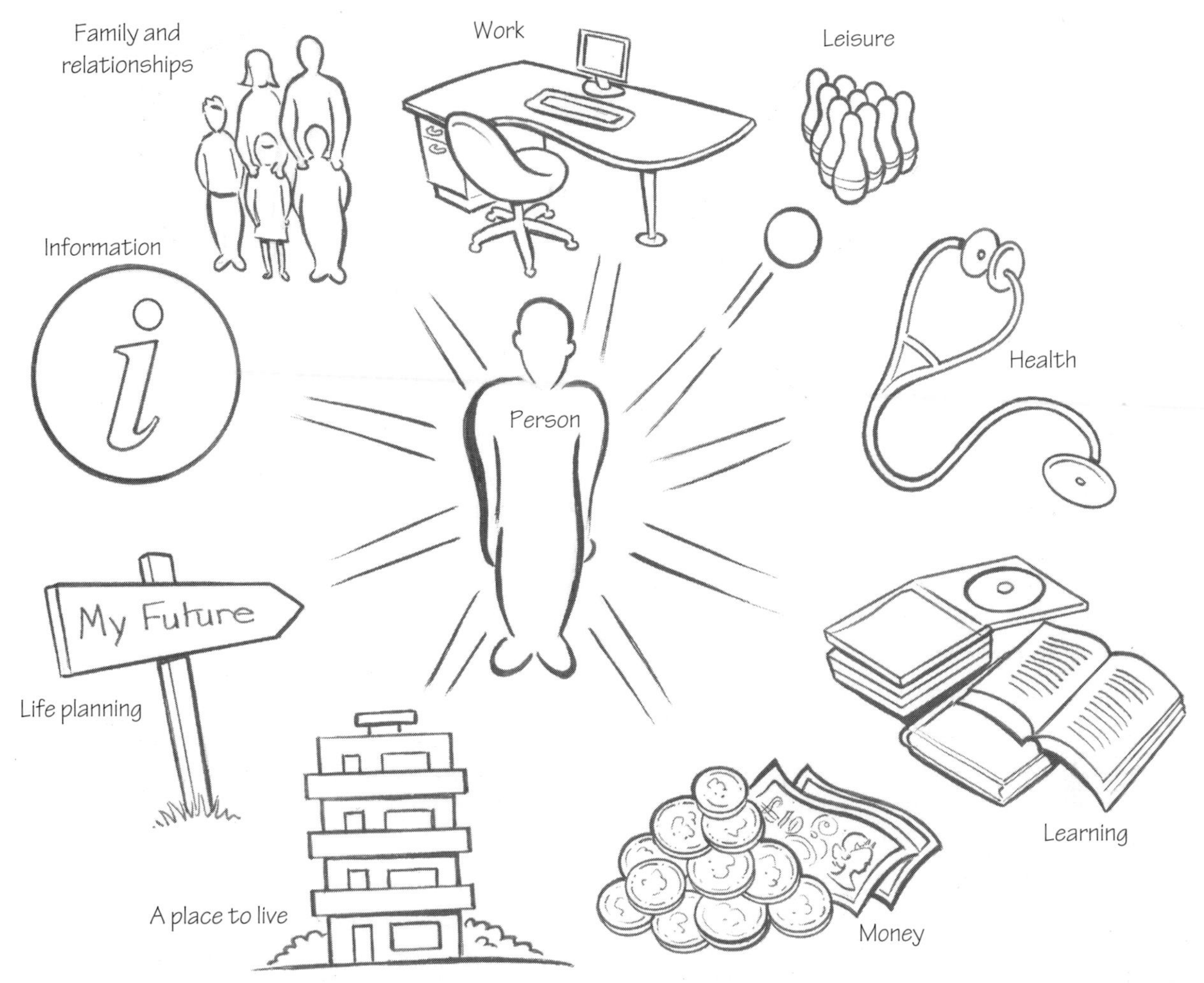

Figure 3.6 PCP means involving service users fully

5. Things do not stop when the first plan is written. Everyone involved keeps listening, learning and making things happen – putting the plan into action helps the person achieve what they want out of life.

(DOH, 2002)

Action on Elder Abuse provides some key questions to consider as part of a person-centred approach:

- What is the person communicating about their views? How can we help them understand and communicate more?
- What is life like for this person? How do they experience the world? What would it be like to be in their shoes?
- What is important for them?
- What might their hopes and dreams be?

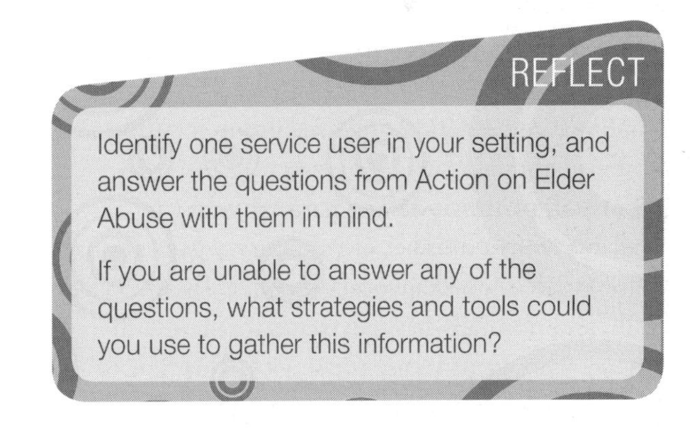

REFLECT

Identify one service user in your setting, and answer the questions from Action on Elder Abuse with them in mind.

If you are unable to answer any of the questions, what strategies and tools could you use to gather this information?

Valuing Medway People Partnership Board and the Circles Network have published some useful examples of tools and formats that can be used in PCP. Some examples are provided below but more can be viewed on their website, and on the Circles Network website.

Advocacy

Advocacy is a process of supporting and enabling people to:

- express their views and concerns
- access information and services
- defend and promote their rights and responsibilities
- explore choices and options.

(Mind, 2008)

Advocacy may take place over a specific issue, or form part of a longer-term relationship between an individual and an advocate. The use of advocates can form a key component of person-centred planning. There are several types of advocacy, some of which are outlined below.

Citizen advocacy matches individuals to advocates who are members of the local community, and who are normally trained volunteers. It is a one-to-one relationship, which may take place on a long-term basis, rather than for one specific issue or area of work.

Peer advocacy is where a service user who has had similar experiences or used similar services assists another to express views and concerns. This may help service users to have a more equal relationship with their advocate.

Group advocacy, also known as collective advocacy, is where a group of people with similar experiences put forward shared views. This could be a national organisation or a local group focusing on local or national issues.

Formal, professional, or paid advocacy is available from a number of voluntary organisations that have developed advocacy services. These services train, and normally pay, advocates to work with anyone who wants to use the service. This type of advocacy is often focused on short-term work.

Independent Mental Capacity Advocacy (IMCA) is for people who do not have the capacity to make their own decisions, and who have no close family or friends to represent their views. IMCAs support people through issues around serious medical treatment or long-term care moves.

REFLECT

- What is your experience of using advocacy services?
- Which type of advocacy is most appropriate in your setting and why?

Care planning and reviews

Care planning and reviewing care plans are key aspects of working with service users and carers. In line with PCP, a theme running through relevant legislation and policies is that the service users' needs must be at the centre of all planning and review processes. However, while person-centred plans created within *Valuing People* guidance (DOH, 2001) should not be concerned with the allocation of resources or eligibility criteria, care plans are underpinned by specific legislation and policy (for example, the NHS and Community Care Act, 1990 and Fair Access to Care Services (DOH, 2002) and do need to consider eligibility criteria and resources.

It can be easy to get caught up in the planning processes themselves, the administration and language associated with them, and to forget their true purpose and focus. This can be seen in the quotes from service users below:

> 'I don't have a care pathway. I have a life.'

(Scottish Executive, 2005:19)

> 'Care plans need to be done properly and people need to be involved. Some people don't even know what a care plan is. When people are not feeling good, they need support to be involved.'

(Beresford, 2005:5)

Risk assessment and management

'When my mother first went into a care home, she asked if she was allowed to go out. The staff assured her that she was – it's her home, not a prison! Of course, they're always careful that they monitor her situation in case she needs assistance' (CSCI, 2005). Risk is a part of daily life, but our attitudes and approaches to risk vary depending on a range of factors such as context, time, situation, previous experience, and so on.

REFLECT

Identify all of the risks you have taken so far today.

- Did you take time to consider the risks or accept them as 'normal'? Why?
- Were the risks 'practical' (for example, taking chances such as standing on a worktop to change a lightbulb) or 'recreational' (for example, 'getting a buzz' from activities such as waterskiing)? (Adapted from Titterton, 2005)

Think of the last time you took a risk that you spent time thinking about.

- What was the risk?
- How did you decide whether to take the risk?
- Did you spent time identifying the pros and cons of taking the risk and if so what were they?

Safety-first approach

This approach:

- tends to focus on physical health and disability, ignoring other needs
- denies rights to choice and determination
- leads to loss of self-esteem and respect
- can lead to a form of institutionalisation, a loss of individuality and an increase in dependence
- at worst, can lead to abuse of vulnerable people.

Your attitude to risk taking is likely to be different if you are considering risks you are taking in your personal life or risks that service users are taking in theirs. This is in part because, as a social care worker, you are bound by law, policies and procedures, including a duty of care. However, Titterton (2005) argues that it is also often because of the stereotypes that professionals have of service users. He argues that professionals tend to view service users primarily as vulnerable people, and to focus on what they cannot do. Titterton (2005) refers to this approach to risk as the 'safety-first approach'.

Finally, Titterton (2005) asks, if we strive for and achieve total safety, does that not result in total control of service users? This is obviously far from a person-centred approach.

The Department of Health agrees with Titterton's view of approaches to risk that focus primarily on safety, and stresses that considering the safest option 'may not be necessarily the best option for the person and may be detrimental to quality of life and a risk to maintaining independence' (DOH, 2007a).'

Risk assessment and management is therefore more than simply risk eradication (if that were possible) or minimisation. It entails efforts to increase potential benefits and to provide a process for planning risk-taking strategies and for monitoring and reviewing the results (Titterton, 2005). Titterton calls this a risk-taking approach.

Risk-taking approach

This approach:

- sees risk as positive, and as enhancing lives
- recognises the importance of psychological and emotional needs as well as physical needs
- promotes choice and autonomy
- values the individual
- promotes rights of vulnerable people and their carers, while accepting that these will sometimes be in conflict.

CASE STUDY

Approaches to risk management

'My friend Chris works all day and can only come and see me in the evening. He asked if this was OK, and the staff assured him that it was fine – 'there are no "visiting times" here' – after all it's my home. They don't just let people traipse in and out though as they have to look after our safety too.'

(CSCI, 2005)

- *Do you think the practice above demonstrates a safety-first or risk-taking approach to visits from relatives and friends?*
- *If you applied the safety-first approach to relatives and friends visiting service users in your setting, what type of procedures and rules would you have in place?*
- *If you applied a risk-taking approach, what types of procedures and rules would you have in place?*

In risk assessment and management, it is necessary to consider the following areas:

- legislation – what is the relevant legislation and how will it guide your decision making and practice in the specific situation you are dealing with?
- policy – which policies apply to the situation?
- procedures – which procedures apply to the situation?
- rights and responsibilities – whose rights and responsibilities need to be safeguarded?
- ethics and values – what are the values issues which need to be considered?

(Cree and Wallace, 2005)

A key area to consider during the planning and review process are the dilemmas that can be encountered when balancing rights with risks. A service user may decide that they want to take part in an activity that could have a serious affect on their health and well-being. As the manager of the setting you will be part of the decision-making process that can restrict what a service user does.

The Care Services Improvement Partnership provide a helpful tool to support decision making.

CASE STUDY

Reviewing risk

Phillip has been a resident in your setting for four years. He has a learning disability and communication difficulties, which can make it quite difficult for him to express his views to others. Phillip has been very independent; he has been working in a local supermarket two days a week, taking himself to and from work. He also enjoys going to rugby matches with a member of his family. His key worker Justin has a good working relationship and communicates well with him. Recently Phillip has started to have 'fits', which are being investigated by the local hospital. This has affected the level of independence he has, impacting on his ability to work and go on social outings, which Phillip is finding very difficult to understand.
You need to review the initial risk assessment regarding Phillip's circumstances and have a review meeting.

Answer these questions.

- *How will you support Phillip during this period?*
- *How will you help to understand the changes to his health, work and social life?*
- *How will you involve Phillip in the risk assessment and review meeting processes?*
- *What do you need to do prior to the review meeting?*
- *Who else will you involve in the review? Why?*
- *What strategies and supports could be put in place to ensure Phillip retains as much independence as possible?*
- *Have you addressed each of Titterton's steps (listed on the following page)?*

Finally, Titterton (2005) outlines a number of key steps which should be involved in risk planning and management:

- Consult and communicate – with all involved.
- Prepare a risk plan – this provides a framework for risk management.
- Ensure 'sign up' – all those involved taking ownership of the risk decision.
- Share information and maintain awareness.
- Monitor and review.
- Support staff.

Managing information

As the manager of a residential setting you will hold and manage a wide range of information about the setting, staff, service users and other professionals and services you come into contact with. When managing information you need to be aware of your settings policies and procedures regarding the recording and storing and confidentiality of records, and the legislation underpinning this. The main piece of legislation to consider is the Data Protection Act 1998.

Data Protection Act 1998

The Data Protection Act came in to force on 1 March 2000. It states that individuals have a right to access records and information about themselves that are held by social services, health and housing organisations. Access to information can be denied if the relevant organisation feels that it will cause serious harm to the individual or another person.

The Act identifies enforceable principles, including these.

- Personal data shall be processed fairly and lawfully.
- Personal data must be obtained only for specified and lawful purposes, and should not be processed in any manner incompatible with that purpose.
- Personal data must be relevant and not excessive in relation to the purpose or purposes for which they are processed.
- Personal data must be accurate and kept up to date.
- Personal data processed for any purpose or purposes shall not be kept for longer than is necessary.
- Appropriate measures must be taken against unauthorised or unlawful processing of personal data and against accidental loss or destruction of, or damage to, personal data.

Service users should also be aware of the nature and amount of information held within your setting that is related to them. Wherever possible, their consent should be sought before information relating to them is shared with others.

There is a great deal of emphasis on professionals working together and sharing information to meet service user needs and provide compatible services. To meet the requirements of the Act, you must ensure that service users are told what information you are going to share with other professionals and why you are going to do this. It is important that information is shared with other professionals on a

REFLECT

Identify the range of people you share information with, and the types of information you need to share with them. Identify the reason(s) why you need to share this information with them.

- What methods of sharing information have been the most effective? Say why you think so.
- What difficulties have you encountered when sharing information? Identify why you encountered the difficulties and what action you took to minimise the problem.
- Were you able to maintain a positive working relationship with the professionals you encountered difficulties with? If so, how did you do this? If not, what could you have done differently to ensure your working relationship was not affected?

need-to-know basis, rather than them having open access to all records.

There can be difficulties when sharing information with other professionals. They may want access to more information than you feel is needed, or than the service user wishes them to have.

CASE STUDY

As already stated, a range of dilemmas may be encountered when sharing, and deciding whether to share, information. Consider the case studies below and identify what action you would take.

Family pressure

A family member pressurises members of your staff team to share personal health information about a service user that the service user doesn't want shared with them.

- *How would you manage this situation with staff, the service user, and family member, ensuring that productive relationships are maintained?*

Shift team problems

It has come to your attention that there are some problems in staff not sharing relevant information and updating other team members fully at shift handovers. This has caused some gaps in continuity of care between shift teams.

- *How would you manage this situation with the shift teams?*
- *What strategies could you adopt to minimise the risk of this situation occurring again?*

This chapter has covered a range of issues related to working with service users and carers. Given the value of person-centred practice that underpins this chapter, it is particularly important to conclude with some quotes from service users reminding us what they want and need from everyone working in health and social care services.

Beresford (2005:27) comments that 'if one word sums up most what service users want from social care, it is "listening"… Listening to and respecting service users was seen as a crucial theme for good social care.'

> 'They need to listen, listen, listen and not patronise us.'
>
> 'They need to listen more to people and take note of what is being said to them. One of the faults tends to be that people in authority just use that authority and do not get involved with the people they are supposed to be helping.'
>
> 'Respect is the major thing in the provision of social care.'
>
> 'An understanding and caring attitude, a bit more of their time, more information of what you're entitled to.'
>
> 'They need to be more supportive – help people work out solutions.'
>
> 'They need a good sense of humour!'

(Beresford, 2005:27,29,30)

NVQ/SVQ

This chapter will help you to provide evidence for the following units:

- **mandatory units A1, B1, C1, E1**
- **optional units A2, B2, B3, B5, B8, E3, E4**

Work Products you may generate and either include in your portfolio or show your assessor to demonstrate your skills and knowledge are:

- **person-centred plans you have been involved in the development of**
- **care plans**
- **minutes of review meetings**
- **risk assessment and management plans you have written**
- **evaluations of areas of service delivery and action plans**
- **minutes of meetings, such as network or interprofessional meetings**
- **evidence of your work with advocates.**

References

Age Concern (2008), www.eurolinkage.org/AgeConcern

Beresford, P. (2005) *Developing Social Care: Service Users Vision for Adult Support*, London: SCIE

Chavez, L. and Guido-DiBrito, M. (1999) *New Directions for Adult and Continuing Education*, San Francisco: Jossey Bass

Cree, V. and Wallace, S. (2005) 'Risk and Protection'. In Adams, R., Dominelli, L. and Payne, M. (2005) *Social Work Futures: Crossing boundaries, transforming practice*, Basingstoke: Palgrave Macmillan

Cross, W. (1995) *In Search of Blackness and Afrocentricity: The Psychology of Black Identity Change*, New York: Routledge

CSCI (2005) *Care Homes for Older People: national minimum standards*, London: CSCI

DOH (2001) *Valuing people: A new strategy for learning disability for the 21st century: planning with people: towards person centred approaches* – accessible guide

DOH (2007a) *Independence, choice and risk: a guide to best practice in supported decision making*, London: DOH

DOH (2008a) '*Direct Payments*' www.dh.gov.uk/en/SocialCare/Socialcarereform/Personalisation/Directpayments/index.htm (accessed 07.09.08)

DOH (2008b) '*Disability*' www.dh.gov.uk/en/SocialCare/Deliveringadultsocialcare/Disability/index.htm (accessed 07.09.08)

DOH (2008d) *Transforming Social Care*, LAC (DH) (2008)

Elder-Woodward, J. (2005) *Factsheet 5: Models of Disability*, Edinburgh: UPDATE Scotland's National Disability Information Service

Holmes, J. (2004) *John Bowlby and Attachment Theory*, Hove: Brunner-Routledge

Innes, A., Macpherson, S. and Mccabe, L. (2006) *Promoting person-centred care at the front line*, York: Joseph Rowntree Foundation/SCIE

Kroger, J. (1997) 'Gender and identity: The intersection of structure, content, and context', *Sex Roles* 36 (11/12): 747–770

Merriam, S. and Caffarella, R. (1999) *Learning in Adulthood: A Comprehensive Guide* (2nd edition), San Francisco: Jossey-Bass

Mind (2008) *The Mind Guide to Advocacy*, London: Mind

Oliver, M. (1990) *The Politics of Disablement*, New York: St Martin's Press

Scottish Executive (2005) *Better Outcomes for Older People*, Edinburgh: Scottish Executive

The Alzheimer's Society (2008) *What Standards of Care Can People Expect From a Care Home?*, www.alzheimers.org.uk/ (accessed 04.06.08)

The Circles Network (2008) *What is Person Centred Planning?* , www.circlesnetwork.org.uk/what_is_person_centred_planning.htm (accessed 04.06.08)

Titterton, M. (2005) *Risk and Risk Taking in Health and Social Welfare* London: Jessica Kingsley

CHAPTER 4

Workforce planning and development

'The social care and social work workforce is the backbone of this country's care system. If we are to ensure that children and adults are looked after in a way that preserves their dignity and peace of mind, it is vital that we have a world class workforce that is skilled, dedicated and valued and supported to do its best.'

(DOH, 2008:1)

'People who use services and carers need to know that those delivering social services bring with them the necessary skills and knowledge to meet their needs. Also that organisations support their workforce in their day to day roles by providing an environment which supports their learning and development throughout their careers. Then we will be able to make sure we have a workforce that is fit for purpose in the 21st century.'

(Scottish Executive, 2005b:2)

Introduction

In this chapter, we begin by considering the concept of a learning organisation and the key processes in enabling an organisation to move towards becoming one. The processes are considered in turn, although each obviously impacts on, and interrelates with, each of the others. The chapter covers:

- performance management
- learning organisations
- workforce planning
- recruitment
- induction
- supervision
- giving feedback
- appraisals
- identifying appropriate learning activities
- exit interviews.

Performance management

The Social Care Institute for Excellence (SCIE, 2005) defines performance management as 'what an organisation does to realise its aspirations and ensure that there are clear lines of accountability for what the organisation does. It is about monitoring performance against targets, identifying opportunities for improvement and delivering change.' SCIE go on to state that effective performance management should demonstrate that:

- you know what you are aiming for
- you know what you have to do to meet your objectives
- you know how to measure progress towards your objectives
- you can detect performance problems and remedy them.

In addition, Armstrong and Baron (2004) stress that performance management is about ensuring that managers themselves are aware of the impact of their own behaviour on the people they manage and that managers should identify and exhibit positive behaviours.

However, balancing these different elements of performance management can be a difficult task to achieve. The focus and implementation of some performance management systems can result in working practices becoming mechanistic and procedural rather than ensuring that social care values are upheld and practice is person-centred.

So, while performance management can help to develop a culture of high performance throughout the organisation and lead to improvements in the standards of service delivered, the systems used to achieve this should be considered within the broader context of the organisation and community it serves. It is vital to ensure that a performance management agenda and systems do not simply result in a target-driven working environment. Managers must always consider whether performance management systems are implemented in a way which maximises the quality of services provided to service users.

Learning organisations

The concept of learning organisations is still relatively new. It has been driven by attempts to identify the key characteristics of successful organisations, and by the need for organisations to manage the rapid changes they have become increasingly subject to.

A learning organisation has been defined as an organisation 'where people continually expand their capacity to create the results they truly desire, where new and expansive patterns of thinking are nurtured, where collective aspiration is set free, and where people are continually learning to see the whole together' (Senge, 1990).

There is some overlap in the concepts of performance management and learning organisations, such as improving performance and the development of all members of the workforce; however, performance management tends to place more emphasis on setting priorities and measuring performance against targets and outcomes, as well as dealing with poor performance. The learning organisation concept emphasises continual learning in all aspects of an organisation's structures and processes much more strongly.

Figure 4.1 The five principal features of a learning organisation

This can be seen in the five principal features of a learning organisation identified by Iles and Sutherland (2001, cited in SCIE, 2004).

1. **Organisational structure** – Managerial hierarchies enhance opportunities for employee, carer and service user involvement in the organisation, including involvement in decision making. Structures support team working, and internal and external networking.
2. **Organisational culture** – The culture promotes openness, creativity and experimentation. It encourages members of the organisation to acquire, process and share information, to innovate, to take risks and to learn from mistakes.

ACTIVITY

Are you a learning team?

Complete the following exercise and consider the extent to which you are a learning team.

	Yes	**No**	**Some/ sometimes**
Are there opportunities for sharing practice issues, e.g. team/staff meetings, group supervision, practice forums?			
Are there processes for accumulating and disseminating knowledge in the team?			
Are there other joint working opportunities within the team?			
Are there other joint working opportunities outside the team?			
Do team members see themselves as responsible for the development of new ideas and methods?			
Are changes planned, managed and reviewed by the team?			
Are all staff involved in the process of leadership or is it only for team and senior managers?			
Is risk taking managed and supported?			
Are mistakes seen as an opportunity for learning, not blame?			
Does the complaints system lead to practice change? How could you demonstrate this?			
Is there any sharing and celebrating of good practice?			
Are new ideas and methods encouraged?			

(SCIE, 2004)

- Have you identified any areas where you have indicated 'No' or 'Some/sometimes'?
- What do you think you could do to improve the areas where you have indicated 'No'?
- Where you have indicated 'Some/sometimes', what factors does this depend on? Do you think you should be achieving 'Yes' consistently in this area, or is it reasonable to achieve this some of the time only? If so, why?
- Where you have indicated 'Yes', identify the factors which help the team to achieve this. Could you apply these lessons to less successful areas?

3. **Information systems** – The systems improve and support practice, and are not used merely for control purposes. The systems facilitate the rapid acquisition, processing and sharing of information.
4. **Human resources practices** – These practices focus on provision and support of individual learning. Appraisal and reward systems are used to measure long-term performance and to promote the acquisition and sharing of new skills and knowledge.
5. **Leadership** – Leaders model the openness, risk taking and reflection that is needed for learning. They effectively communicate a convincing vision for the organisation and provide support to others in striving to achieve the vision. They make sure the organisation and teams within it have the capacity to learn, change and develop.

Some authors believe the concept of a 'learning organisation' is too idealistic and that it is more achievable for organisations to aim to develop and maintain a learning culture, similar to the description of 'organisational culture' outlined in point 2 above.

The National Institute of Social Work (1996) describes managers as having a key role in 'fostering a learning culture, negotiating learning strategies, building learning resources in the workbase, setting an example as learners and acting as facilitators.'

Whichever concepts and terminology you feel most comfortable with you might find it useful to consider the extent to which your team could be considered a 'learning team'. The Social Care Institute for Excellence (SCIE) has produced a *Learning Organisation Self-Assessment Resource Pack* (SCIE, 2004). The activity on page 47 is adapted from that resource, and you might also find it useful to look at other information within the pack.

Workforce planning

Workforce planning is an element of good management practice that has attracted increased attention in the health and social care sector in recent years. SCIE (2007) states that workforce planning is about having the right people, with the right skills, in the right jobs, at the right time. It says that workforce planning – also sometimes called human resource planning – is about addressing two basic questions in terms of your future workforce:

- how many staff?
- what sort of staff?

Workforce planning has achieved a higher profile recently in part because of the significant workforce challenges faced by the sector.

- Turnover rates in social care remain high. It is vital to reduce turnover rates given the difficulties encountered in initial recruitment of staff, and the impact of staff turnover on the quality of service delivery.
- The workforce is ageing. The demographics of the population as a whole are changing, with an increasing proportion of people over 65 years old. As the current workforce retires, it is vital to attract new people into social care as a career. The changing demographics of the population also mean that the demand for social care services will continue to increase (DOH and DFES, 2006).
- There is a skills shortage within the workforce. The DOH continues to report on the low skill levels in the workforce relative to training needs (DOH, 2007), as well as a relatively low level of staff holding a qualification relevant to their role.
- There are ongoing changes in service provision requiring new types of worker and/or workers with different skill sets (see box).
- New legislation and policy may require changes in knowledge, skills and practice. This has included the statutory requirements for qualifications in some job roles.

Despite the challenges faced by the social care workforce, in a national survey of care workers (Skills for Care and CWDC, 2007), 88 per cent of respondents said they were happy in their current job. Reasons given included enjoying the work overall, enjoying working with and helping to improve the quality of life of service users, and liking the team they were working with.

1. In a number of residential care homes in Bath, Bristol and South Gloucestershire, staff have been trained by health staff to carry out some activities formerly carried out by district nurses/health visitors. The training has been closely linked to existing occupational standards and NVQ/SVQ elements. In an independent evaluation of this work, some of the benefits were:

- residents and their families reported an increase of satisfaction with the more 'holistic approach'
- staff reported an increase in job satisfaction, morale and confidence
- links and understanding between health and social care have been improved and better working relationships reported.

2. A research project on the work of Jewish Care highlighted some staff development needs in relation to dementia care. Skills for Care funded three new posts – a Project Manager and two full-time Practice Development Workers for two years – leading to the development and introduction of a systematic competency-based learning and development programme. The project demonstrates the benefits of a systematic and in-depth training strategy, which has led to people working differently. The role of a dementia care coordinator has also been successfully developed.

Other examples of new types of worker/new types of working can be found at www.newtypesofworker.co.uk.

The workforce planning process

SCIE (2005) identifies five stages within the workforce planning process.

1. **Audit of the current workforce** It is important to hold up-to-date, accurate and comprehensive information about your existing workforce. This will help you to identify future needs (for example, arising from a number of staff due for retirement at a similar time) and trends, or particular areas of difficulty, such as high turnover of staff in certain areas or roles. It is also important to consider the diversity of your workforce. The Topss England *Workforce Planning Toolkit* (2004) poses some useful questions for employers to consider including:
   - Does your organisation reflect the diversity of the service users?
   - Are your work regimes flexible enough to encourage people with religious or personal needs for specific time off to work for you?
2. **Consideration of context** You will also need to consider the wider context of the service, including strategic/business plans. You may consider questions such as:
   - Is the business expanding or specialising?
   - Are the needs of service users changing?
   - Are new regulations coming into force that will have an impact on the workforce?
3. **Forecasting** This involves considering the outcome of stages one and two of the process and identifying gaps between the current workforce and what will be needed in the future. This involves identifying not just the numbers of staff but also the skills, knowledge and qualifications that will be required.
4. **Planning** Having carried out the first three stages, it should become possible to develop a plan for the future. SCIE (2005) identifies three areas which might be covered within these plans:
   - recruitment and succession/career planning
   - workforce development programmes to support career development and retention of staff
   - changes to reward/benefits packages for staff.
5. **Implementation** The workforce plan should be closely and clearly linked to the organisation's strategic/business plan, as well as unit and/or team plans, and should be reviewed in line with these other plans.

Workforce planning is not an easy process, but registered services are required to show evidence of

ACTIVITY

Workforce planning

Answer the questions below to begin to identify future workforce needs.

1. Is your service area planning any major growth or downsizing over the next one, two or three years, and what impact will this have on the current workforce?
2. Is the way in which you provide your service likely to change over the next one, two or three years: for example, via increased use of technology or more flexible provision of service. If so, what would be the impact on your workforce?
3. Have you factored into your service and workforce planning the impact if sickness absence levels remain at their current level, or have you developed and implemented action plans for reducing absence levels?
4. Have you factored into your service and workforce planning the impact if turnover levels remain at their current level, or have you developed and implemented action plans for managing these turnover levels?
5. What is the age profile of your current workforce and what is the likely impact of retirements over the next one, two or three years?
6. Does the diversity profile of your workforce reflect the local community and is there any action you could take to improve this?
7. Are there any legislative, regulatory or other changes expected over the next one, two or three years that will have an impact on your service area, the size of workforce required or the learning needs of the workforce?

(Adapted from London Councils, no date)

- What do you anticipate the impact of each of these issues will be on your workforce, in terms of workforce numbers, structures and required skills?
- What strategies could you put in place to manage the impact of each of these issues?

workforce planning (CSCI, 2006). You may find the Topss *Toolkit* (2004, 2nd Edition) useful, particularly for the first three stages of the process. It is also advisable to concentrate on simple, focused plans that are reviewed regularly, rather than an over-ambitious plan.

Recruitment

NVQ/SVQ

This section will help you to provide evidence for units A3 and B1.

Staff are the most valuable resource that any setting has. It is much more cost-effective to invest time at the beginning in getting the right person for the right job than dealing with poor performance at a later stage. Briggs (1976, cited in Collins, 2006) stresses this point saying that it is important to 'first get the right people on the bus …' because it is 'more difficult to the wrong people off the bus' (Collins, 2006:13-14). Briggs argues that employing 'great' staff should be the priority if you are going to provide a quality service. However, the CSCI (2006:3) identify the recruitment and vetting of care staff as one of the 'poorest areas of performance against the National Minimum Standards in regulated social care services in England'.

CSCI (2006b) remind us that 'recruitment and vetting practices have two vital functions in social care. These are: first and foremost, to protect children and adults by ensuring that the people who provide their care are suitable to do so and treat them with dignity and respect; and secondly, to protect the rights of applicants to be considered equally for vacant posts'.

When recruiting staff it is important that you know, understand and follow your organisation's recruitment procedures. Most organisations provide training for all those taking part in the recruitment and selection of staff. This ensures that all those concerned are aware and comply with legal and industry requirements.

There is a range of legislation that relates to, and regulates, recruitment. Some of the main Acts and regulations are listed here.

Legislation	Relevance to recruitment
Rehabilitation of Offenders Act 1974	The Rehabilitation of Offenders Act concerns the employment of people with a criminal record. Where the type of work to be undertaken will bring the person in contact with vulnerable groups, e.g. the disabled, the elderly, mentally ill and young people under 18 years of age, those with a criminal record must disclose the details of all their convictions including those which are 'spent'.
Sex Discrimination Act 1975	It is unlawful to discriminate in employment, including recruitment, on the grounds of a person's sex or marital status.
Race Relations Act 1976	It is unlawful to discriminate in employment, including recruitment, on the grounds of a person's race, colour, nationality, ethnic or national origin.
Disability Discrimination Act 1995	The Act seeks to ensure that a disability should not bar a person from employment unless it would genuinely prevent them from doing the job and there is nothing the employer can reasonably do to overcome difficulties resulting from the candidate's disability. In the recruitment process it is important to ask individuals whether they require any special arrangements to be made in order for them to participate fully.
Asylum and Immigration Act 1996	An employer could be guilty of a criminal offence if they employ someone who does not have permission to work in the UK.
Data Protection Act 1998	Information (in any form, e.g. written, electronic) must be collected and used fairly, stored safely and not disclosed to anyone unlawfully.
Protection of Children Act 1999	It an offence for any organisation to offer employment to anyone who has been convicted of certain specified offences, or is included on lists of people considered unsuitable for such work held by the Department for Education and Skills and the Department of Health, where the work involves regular contact with children. It is also an offence for people convicted of such offences to apply for work with children.
Care Standards Act 2000	The Act created the Protection of Vulnerable Adults (POVA) list. Individuals' names are included on the list if they have abused, neglected or otherwise harmed vulnerable adults whether or not in the course of employment; acting as a workforce ban. POVA checks are requested from the Criminal Records Bureau.
The Safeguarding Vulnerable Groups Act 2006 and the Protection of Vulnerable Groups (Scotland) Act 2007	This law will creates two, linked, lists of people barred from working in 'regulated activities' with either young people or vulnerable adults. A new organisation, the Independent Safeguarding Authority (ISA), has been established with the main aim of preventing unsuitable people from working with children and vulnerable adults. It will do this by placing these people on one of two ISA Barred Lists, and making decisions about who should be on these lists. The ISA is due to 'go live' on 12 October 2009.
Part-Time Workers (Prevention of Less Favourable Treatment) Regulation 2000	The Regulations make it unlawful for part-time workers to be treated less favourably than full-time workers. In terms of the recruitment and selection process, care should be taken to offer the same terms and conditions to all employees regardless of their part- full-time status.
Employment Equality (Sexual Orientation) Regulations 2003	The regulations came into force on 1 December 2003. They provide protection in the workplace against all forms of discrimination based on sexual orientation. In terms of recruitment and selection processes, care should be taken not to ask personal questions, which may be perceived to be intrusive and imply potential discrimination.

Legislation	Relevance to recruitment
Employment Equality (Religion or Belief) Regulations 2003	The regulations came into force on 2 December 2003. They provide protection in the workforce against all forms of discrimination based on grounds of religion or belief. ACAS (Advisory, Conciliation and Arbitration Service) advise that 'where it is reasonable to do so, organisations should adapt their methods of recruitment so that anyone who is suitably qualified can apply and attend for selection. Some flexibility around interview/selection times allowing avoidance of significant religious times (for example, Friday afternoons) is good practice.' (2005:p13)
Employment Equality (Age) Regulations 2006	The regulations came into force on 1 October 2006. They provide protection in the workforce against age discrimination. It is now unlawful for an employer to impose a lower age limit when recruiting, unless this age restriction can be objectively justified or is imposed by law.

Here is a flow chart of the recruitment process. Some elements are then considered in further detail.

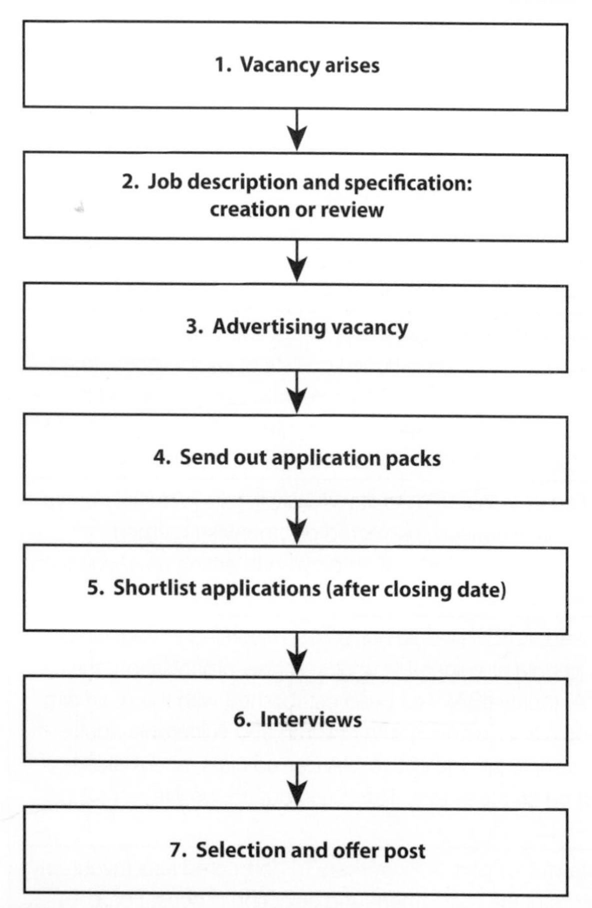

Figure 4.2 The recruitment process

Vacancy

A vacancy may arise due to staff turnover, the changing needs of the organisation and/or service users, or as a result of a reorganisation or secondment.

Job description and person specification

An important part of the recruitment process is to prepare a job description and person specification. The job description should be an objective and unbiased, formal analysis of the role, including the responsibilities of the postholder and the goals and objectives of the role. The person specification is a profile of the person you consider best fits this analysis, and will outline the qualifications, experience, skills and knowledge an individual will need to have in order to carry out the job role and responsibilities. The criteria will be specific and should be able to be evidenced by the applicant. This will form the basis of the decision whether an applicant should be shortlisted for the advertised position.

Job descriptions and specifications also need to take into account any National Minimum Standards or National Occupational Standards that apply to the role. It is important to be aware of developments and changes to these standards, and to review the relevant websites regularly.

For a job description to be a useful and realistic tool, it should be reviewed and developed as job roles change, and when vacancies are to be advertised. This can be quite difficult in organisations where there are generic job descriptions for all staff at the same level. In an NVQ/SVQ Management workshop, an exercise was carried out with the group of managers, who all worked for the same organisations, but in different settings, which identified how different their job roles were. Each of the managers carried out an analysis of their job description in relation to their actual duties, and the exercise clearly highlighted that job descriptions did not match the current role or the level of responsibility that they each had.

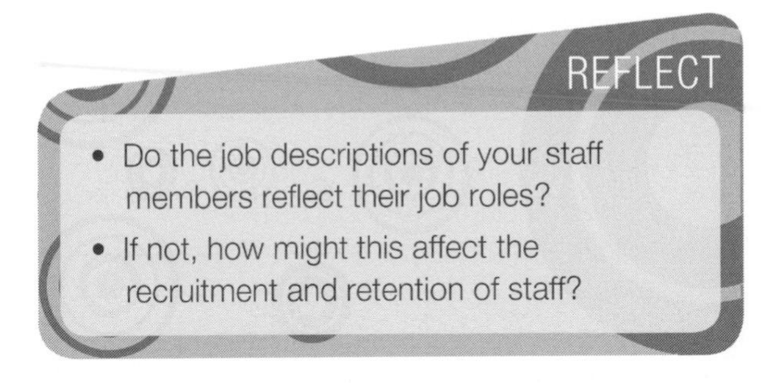

Application packs

As well as the application form, packs sent to applicants will normally include:

- information about the organisation
- job description and person specification
- information about salary
- Equal Opportunities statement
- confirmation that the successful candidate will be subject to Criminal Records Bureau (CRB) and Independent Safeguarding Authority checks.

Shortlisting

The purpose of shortlisting is to identify those applicants who best meet the job specification for

Job title / Reference number	Applicant 1 Name / Ref. no.	Applicant 2 Name / Ref. no.	Applicant 3 Name / Ref. no.	Applicant 4 Name / Ref. no.
Criteria	Rating 1–5	Rating 1–5	Rating 1–5	Rating 1–5
Essential criteria 1				
Essential criteria 2				
Essential criteria 3				
Essential criteria 4				
Desirable criteria 1				
Desirable criteria 2				
Desirable criteria 3				
Total scores				
Completed by:			Date:	

Scoring guide:
5 = Meets fully, 4 = Meets most of the requirement, 3 = Just meets the requirement, 2 = Just fails to meet the requirement, 1 = Not met

Figure 4.3 A sample shortlisting record

the post, and therefore warrant further consideration (see Figure 4.3). Shortlisting should be based on the extent to which the application addresses the requirements of the job description and person specification, and no other criteria.

References may be taken before the interview. If this is not possible, any offer of employment **must** be made conditional on suitable references, as well as a range of other evidence (*see Selection and offering the post*, below). A minimum of two references should be taken up, and any concerns, gaps or ambiguous statements followed up directly with the referee. In all situations, it is good practice to speak with one of the referees directly. Any gaps in employment should be noted and explored fully, during the selection process and before any offer of employment.

Interviews

There should be a minimum of two people on the interview panel. The same, pre-prepared questions should be asked of each interviewee, and notes taken of their responses. It is acceptable to ask follow-up questions to gather more information from the interviewee. Interview questions should relate to the job description and person specification, and should enable the interviewee to demonstrate how they meet them.

You may also wish to consider whether and how to involve service users, carers and/or other stakeholders in the interview process. This could be as an additional member of the interview panel, as part of a group interview, or in other parts of the selection process.

Townsley et al. (2002) undertook a project aimed to promote and support the involvement of people with learning difficulties in staff recruitment. You may find it useful to consider their work if you wish to develop this aspect of your recruitment and selection processes. Details are provided in the Useful reading and websites section at the end of this chapter.

CASE STUDY

Interviewing

You are taking part in an interview panel with another manager from your organisation, recruiting an assistant carer for the residential home she manages. Your organisation has a set of interview questions for this job role that have been used for the last two years.

During one interview, you become aware that the applicant seems to anticipate the questions you are about to ask, and answers each question covering the elements you have in the organisation notes for the 'ideal response'. You are aware that the applicant has a friend who already works for the organisation.

- *Why would this situation give you cause for concern?*
- *What action might you take? Who else in the organisation could you seek guidance from?*
- *How could this type of situation be prevented in the future?*

Selection and offering the post

A conditional offer of employment must not be made until the following have been checked.

- **Identity and permission to work** – This should include name, date of birth, address and photographic identity. Permission to work refers to the requirements of the Asylum and Immigration Act (1996) that a person has permission to work in this country.
- **Qualifications** – Original copies of all relevant qualifications must be seen and copies retained on file.
- **CRB clearance** – It is important to note that criminal record information is only supplied for the purpose of making a recruitment (or other

relevant) decision. Once this has been made, the information must be disposed of securely unless, in exceptional circumstances, the information is clearly relevant to the ongoing employment relationship (Criminal Records Bureau, 2007). You should seek further advice where you consider this to be the case.

- **Barring list clearance** – The new Independent Safeguarding Authority (ISA) brings together the existing barring schemes – Protection of Vulnerable Adults (POVA), Protection of Children Act and List 99 – and will support the implementation of the Safeguarding Vulnerable Groups Act 2006. It will draw on wider sources of information to provide a more comprehensive and consistent measure of protection across a wide range of settings, including the whole of social care and the NHS.
- **Registration with regulatory bodies** – such as General Social Care Council or Health Professions Council (where required). Original copies of registration documents should be seen and copied, but it is also vital that employers check that the registration remains current. It is possible that the individual may have been subject to a conduct hearing during the period of registration. This situation may not come to your attention until the individual is due to re-register with their professional body. So it is important for employers to check registration details online at the professional body's website (see Figure 4.4).
- **Health and fitness** – Where the individual is not required to be registered with a professional body, it may still be necessary to confirm that the individual has the appropriate level of health and fitness to carry out the duties of the post.

  Remember that under the Disability Discrimination Act, 2005, employers are required to make 'reasonable adjustments' to ensure that they do not discriminate against employees, including potential employees. This might mean making changes to the working environment to make it more accessible, or providing adapted/specialist equipment for aspects of the individuals work (such as adapted computer equipment).
- **Satisfactory References** – See the section on Shortlisting above regarding satisfactory references.

Records to keep on file

The registration regulations set out the documents to be kept on file for each employee. They normally include:

- name, address and date of birth
- proof of identity (for example, photographic identification, copy of passport and/or birth certificate). Applicants should provide details of any other names that they may have been known by: maiden names, names changed by deed poll and 'known as' names, as checks may need to be made against those names.
- CRB disclosures
- two written references, dated and addressed to the prospective employer
- written verification of reasons for leaving previous similar employment (if available)
- full employment history
- statement by the employee of mental and physical fitness
- details of registration with any professional body
- description of duties and responsibilities.

The CSCI (2006) recommends that the following documents are also kept on file:

- completed application form
- evidence of the interview format, and the questions and answers given by the candidate
- evidence of induction, probation periods and supervision
- contract of employment
- signed confirmation of the candidate's receipt of their terms and conditions and copy of the General Social Care Council's Code of Practice (2002).

General **Social Care** Council

Home

SOCIAL CARE STANDARDS...

Codes of practice

The Social Care Register

- The Social Care Register explained
- Protection of title
- Apply for registration
- Post-registration
- Registration processing times
- **Check the register**
- Transfer and additional country registration

Conduct

TRAINING AND LEARNING...

Training and learning

For course providers

OUR ORGANISATION...

News and events

About us

Contact us

Working for us

Links

Check the register

Search the Social Care Register

You can use this page to check a social workers registration. You can search by name, town where the social care worker is currently employed (e.g. London), or by social care registration number. To refine your search results you just fill in more than one field.

The absence of a record does not necessarily mean that the worker is not registered. Just call us if you do not find the worker you are looking for. You can only search the register for England on this website – use one of the links on the right to search Scotland, Wales or Northern Ireland.

First name(s)

Surname

Town of employment / study (e.g. London)

Social care reference no.

Part of the Register: Please Select...

Submit

SEARCH

OTHER PAGES OF INTEREST

The Social Care Register explained

HELPFUL WEBSITES

Social Care Register for Scotland

Social Care Register for Wales

Social Care Register for Nothern Ireland

NEED A HAND?

Registration helpline: 0845 070 0630

email: registration@gscc.org.uk

Disclaimer | Copyright notice | Privacy policy

Access key information | Site map

hpCheck.org

Be sure they're registered

Who are the HPC & what do we do? Why check your health professional is registered? What does registration mean?

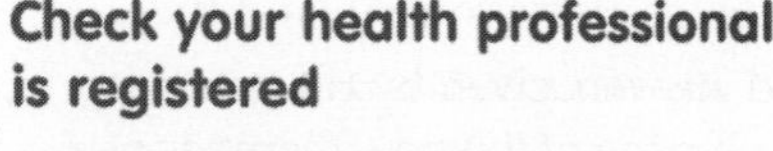

Check your health professional is registered

We are the Health Professions Council, and we were set up to protect your health and wellbeing. To do this, we keep a register of health professionals who meet our standards for their training, professional skills, behaviour and health.

You can use the on-line register on this page to check if your health professional or the health professional you are about to use is registered.

Who do we regulate?

View our main site

Step 1: Select a profession

...Please Select...

Step 2: Enter a Registration Number OR Surname

Registration Number:

Surname:

Need assistance?

Search

Home | Email page | Print page

Park House, 184 Kennington Park Road, London, SE11 4BU, +44 (0) 20 7582 0866

Who are the HPC and what do we do? | Why check your health professional is registered? | What does registration mean?

Design and Technology by Reading Room

hpc Health Professions Council

Figure 4.4 Checking professional registration

VIEWPOINT

- Do you know your organisation's recruitment procedures?
- Are you clear about your role in the various stages of recruitment?
- Are you fully aware of the legal requirements to be considered when employing staff?

Induction

'Effective induction is seen as a vital component of providing safe care and of establishing a competent workforce.'
(Skills for Care and CWDC, 2007)

In the past, induction has not always been seen as an area of importance, but more a 'tick list', to complete as quickly as possible. With the introduction of the Induction Standards by Skills for Care in 2001, greater value was placed on the induction process, with managers of settings taking more responsibility for ensuring its effectiveness. The Induction Standards cover areas such as health and safety in the workplace, policies and procedures, and staff understanding their roles and responsibilities.

Welcoming staff is the first stage of induction. Beverly and Coleman (2005) identify a number of ways to make new people feel welcome, including:

- sending directions on how to get to the setting
- ensuring people are met promptly on arrival
- including other colleagues in some of the induction process
- ensuring people have time to settle in and orientate themselves with the service
- providing initial information about service, resources and policies
- providing a specific person to whom they can relate when you are not available
- providing information about the setting's protocol: for example, how does the tea fund work?

REFLECT

Reflect on how you make new members of staff feel welcome.

- What do you do before they start?
- What do you do on their first day?
- Do you include areas raised by Beverly and Coleman, as above?
- What else would you include on your list? Why?

CASE STUDY

Offering new challenges

Amina has worked in your setting for five years; she has completed her NVQ/SVQ2 and NVQ/SVQ3 during this time. She is key worker for two service users. She knows and understands the service users' needs and works well with them. She has been a caring, committed member of your team who has taken on additional responsibilities and is able to use her own initiative.

In her recent supervision, she told you that she was not enjoying her job as much as there are no new challenges and as a result she is feeling demotivated.

You have a new member of staff starting in your setting. The staff member is also new to residential work.

- *How could you include Amina in the induction process of the new member of staff?*
- *How might this help the new member of staff?*
- *How might this benefit Amina?*
- *How might Amina's involvement in the induction benefit service users and the setting overall?*

Induction can be enhanced by including experienced members of staff. This can benefit the staff team and setting, as it enables others within the team to share their skills and knowledge. Experienced members of staff can lack motivation if they are not given new challenges. Including them in the induction process can be good way to make them feel involved and valued, and also enable them to develop new skills, especially if the member of staff concerned wants to develop supervisory skills.

Induction can be used to help staff make links to NVQ/SVQs and to begin to gather some of the evidence they will need to meet the NVQ/SVQ requirements. For example, as they are completing sections on health and safety they will be able to use these for either unit HSC22 or unit HSC32, depending on the level of their qualification. The more support staff are given at the induction stage, the more they will understand and feel confident about NVQ/SVQs and the more likely they will be to achieve the qualification.

Induction standards for managers

Induction standards for managers have also been developed. These are for experienced managers new to a setting, those who are new to management and those who are potential, or aspire to be, managers. Adult social care management induction standards have been developed by Skills for Care (2008) to support managers during their induction period and to develop areas of their practice and knowledge. Similar to other induction standards, to fully complete the units managers will have to demonstrate their skills and practice in the workplace. As a result there will be some aspects that staff who are not yet managers will not be able to complete as units will require evidence of actual management experience.

There are six standards:

- Standard 1 – understand the importance of promoting social care principles and values
- Standard 2 – providing and facilitating change
- Standard 3 – working with people
- Standard 4 – using resources
- Standard 5 – achieving outcomes
- Standard 6 – managing self and personal skills.

As with other induction standards those completing the management standards will be able to make links to NVQ/SVQ4 units, both within the Health and Social Care award and Leadership and Management. A full copy of the standards can be found in Appendix 3.

REFLECT

Reflect on your induction when you became part of the management team.

- Do you feel that all the areas you needed to know and understand were included in your induction?
- Identify one area that you felt well supported in and one area that you felt you needed additional support with – say why and how.
- How could the management standards have supported you during your induction?

CASE STUDY

Planning inductions

You have recently appointed two new deputy managers within your care home. Jessie has been working in your setting for five years and has been acting deputy for the last six months. Robert is new to your setting, organisation and management. He has worked in residential settings for ten years. He was a key worker and shift leader in the setting he worked in previously. You are responsible for planning their induction and taking them through the management induction standards.

- *Will you plan the same induction for Jessie and Robert?*
- *If yes explain why, if no why not.*
- *How will you incorporate the induction standards?*
- *What are the benefits of having induction standards? Are there any disadvantages?*
- *How can Jessie and Robert support each other during this process?*

Supervision

Brown and Bourne (1996) identify supervision as 'the primary means by which an agency-designated supervisor enables staff, individually and collectively, and ensures standards of practice. The aim is to enable the supervisee(s) to carry out their work…as effectively as possible.'

Thompson (2006) suggests that supervision has four functions:

- an **executive function**, which relates to the role of the supervisor in ensuring that policies and procedures are followed, and the required standards of practice maintained. This element ensures that the supervisee is held accountable for their work
- an **educational function**, which relates to the learning needs of the supervisee
- a **support function**, which relates to care for staff, ensuring they feel supported and valued, and that practice and achievements are recognised
- a **mediation function**, which relates to the role supervisors can play in mediating between the supervisee and senior managers/wider organisation. The supervisor may also feel that they act as the buffer between the two, in situations of conflict.

Not only is it good practice to provide regular supervision, but the GSCC Code of Practice for Employers (2002) and the National Minimum Standards (DOH, 2003a) require that supervision is provided to all social care staff. The National Minimum Standards also set requirements regarding the frequency and content of supervision. In the *National Minimum Standards: Care Homes for Adults (18–65)* (DOH, 2003a), the relevant standard is as follows:

> **Standard 36** Supervision and Support Outcome – Service users benefit from well supported and supervised staff.
>
> 36.4 Staff have regular, recorded supervision meetings at least six times a year with their senior/manager in addition to regular contact on day-to-day practice (fortnightly where there is no regular contact; pro rata for part-time staff), covering:
>
> i. translation of the home's philosophy and aims into work with individuals
> ii. monitoring of work with individual service users
> iii. support and professional guidance
> iv. identification of training and development needs.

Similar practice is required in the *National Minimum Standards: Care Homes for Older People* (DOH, 2003b) (Standard 36).

The standards outline the minimum number of supervision sessions that should take place in a year. There may be some staff in your team who will require supervision more frequently. In addition, the GSCC Code of Practice (2002) refers to the need for employers to ensure they are 'effectively managing and supervising staff to support effective practice and good conduct and supporting staff to address deficiencies in their performance'. This may well require levels of supervision above the minimum required within the National Standards.

Good practice in supervision

In an NVQ/SVQ workshop, a group of residential managers identified a number of elements which they felt made supervision effective and the skills they needed to practice with the supervisee.

- Have an agreement with the staff member which identifies time, frequency, agenda and boundaries.
- Keep clear and accurate records.
- Know and understand the organisation's policies and procedures.
- Be approachable.
- Be organised and prepared.
- Supervision is a two-way process, but remember this is the staff member's time
- Do not be confrontational.
- No surprises (discuss difficulties as they occur; do not save them for supervision).
- Do not feel you have to have all the answers.

- Respect confidentiality, informing staff when and why this cannot be respected.
- Make sure the space supervision is held in is private and that there are no interruptions.
- Keep the seating informal: for example, not sitting behind a desk.
- Help staff link skills and knowledge to practice.
- Give constructive feedback, identifying areas of good practice and areas of development.

ACTIVITY

Good practice in supervision

Look at the list of elements for good practice in supervision.

- Do you agree with the above list?
- Is there anything that you would add or take out – and if so, why would you do this?
- What aspects would you find easiest and hardest to do as a manager?
- Why do you think you find certain areas more difficult? How could you develop your skills, knowledge and confidence in these areas?

Supervision agreements and agendas

Supervision agreements can help when devising the content and structure of the sessions. It also enables the staff member to feel part of the process and ensure the time allocated is not only used to look at current workload.

Thompson (2006:75) says that 'some people unfortunately adopt a narrow view of supervision and see it primarily or even exclusively as a means of ensuring that sufficient quality and quantity of work is carried out … A broader view of supervision can play a significant role in promoting learning and developing a culture of continuous professional development.'

Agreements serve to address a number of areas, such as:

- to set the boundaries and the tone of the sessions
- to identify roles and responsibilities
- to highlight the frequency, length and venue for supervision
- to confirm that supervision will not be interrupted unless there is an 'emergency'. An agreement can outline what both you and the supervisee consider an emergency of this nature to be
- to agree how agenda items will be negotiated
- to confirm the boundaries of confidentiality, including areas that can not stay confidential
- to confirm appropriate reasons and processes for cancellation of supervision.

Agreements also demonstrate that there is a commitment to supervision, and that those involved see it as a valuable and important process.

When developing agreements with staff, it is important that you adapt your practice to meet the needs of the individual staff member. Different approaches will be needed if all staff are to feel part of the process. You may wish to consider whether the content of the agreement would be different for the new member of staff, or for a member of staff in a management position. An example of a supervision agreement is provided on page 61.

REFLECT

- What are your setting's policy and procedures on supervision?
- What is your role in putting them into practice?
- How do you evaluate the effectiveness of the policy and procedures, and your practice in implementing them?

One of the case studies on page 62 gives you the opportunity to think through how you will set agendas with different members of staff. Supervision normally begins with a 'temperature check', giving the staff member an opportunity to tell you how they are. Other items which may be included in an agenda are:

- workload management
- reviewing work plans

Appendix C. Supervision agreement example

Walsall Council

Walsall Metropolitan Borough Council

Supervision agreement proforma

Between: ____________ **and** ____________

Frequency: ____________

Location: ____________

Duration of session: ____________

All information between supervisor and supervisee will be treated with respect and in a professional manner. Supervision can be individual or as part of a group.

Agenda and structure

Formal supervision sessions should be structured with preparation work having been carried out by both the supervisor and the supervisee and, where possible, an agenda set before the supervision session. Any major issues requiring detailed discussion should be put in writing and distributed a few days before the supervision session. Both parties should prioritise the agenda items at the beginning of the session in order to make the most effective use of time. Formal supervision sessions should ordinarily last for about & probably no more than, one to one and a half hours.

Content

Supervision will cover:

- performance management and administrative functions
- learning and development and teaching functions
- supportive functions.

Anti-oppressive

Supervision should be based on anti-oppressive principles and should be sensitive to race, gender, disability, impairment, age, religion and sexuality.

Record keeping

All supervision sessions should be recorded including areas covered, discussion points, agreed action plans, timescales and who the action is to undertaken by. Copies of the record should be available to both supervisor and the supervisee and can be accessed by the supervisor's manager or any other person with a reason to access the supervision record as deemed necessary by the authority's code of conduct.

Cancellations

In the event that a scheduled supervision session has to be cancelled by either party, it will be re-scheduled at the point of cancellation proving to be unavoidable. The session should be re-scheduled to take place within 5 workings days of the date of the original booked session. If the cause of the cancellation is the sickness absence of either party then another supervision session will be booked within 5 working days of the person's return to work. In the event that the supervisor is absent from work for more than two weeks unplanned leave, it is the responsibility of the supervisee to report to the supervisor's line manager for alternative supervision arrangements to be made.

Disagreements

Areas of disagreement between supervisor and supervises will be recorded on the supervision records. Areas of disagreement that cannot be resolved may be referred to the line manager.

Review of supervision

Supervision session – process, content, length, frequency, format and style should reviewed by the supervisor and the supervisee at least annually.

Agreement

We agree that supervision will be given and received in accordance with the Walsall Metropolitan Borough Council Social Care & Supported Housing Supervision Policy wherein more details regarding supervision can be located.

Providing effective supervision workforce development guide

Figure 4.5 Supervision agreement example

- learning and development, including reflection on practice
- arrangements for annual leave and time off in lieu (TOIL)
- team issues
- other issues, such as health and safety.

Developing an agenda is not enough on its own to make a supervision session run smoothly. You still need to take into account the differing personalities, motivations and behaviours of your team members.

Chapter 5 will help you to reflect on some of these behaviours more fully and how you might manage them. In addition, as a manager you need to be aware how you respond to different people and situations as this makes it easier for you to deal with them.

CASE STUDY

Planning for supervision

- *Consider the following members of staff and answer the questions below for each of them.*
  - *Staff Member 1 is a new member of staff who is not very confident and lacks initiative.*
  - *Staff Member 2 uses supervision to gossip about others in the team.*
  - *Staff Member 3 uses supervision only to discuss their personal issues.*
  - *Staff Member 4 does not value supervision and sees it as a waste of their time.*
- *How are you going to work with each of the staff members to make supervision productive?*
- *How will you set agendas and keep the supervision focused?*
- *Will you have common areas on each agenda? Why/why not?*

Recording supervision

As part of the supervision agreement, you can decide who will record the content of the supervision and

CASE STUDY

Dealing with undermining behaviour

You have recently started supervising a member of staff who has worked in your organisation for the same length of time as you. You both started at the same grade, but you have chosen to pursue a career in management, while she has remained at the same grade. She does not feel that she needs to be supervised and has made it clear that you 'can't teach her anything'. You have held two supervision sessions with her, where she has attempted to undermine your role, and has criticised any comments and suggestions you make.

- *Why do you think she may be acting in this manner?*
- *What strategies can you use to make supervision more constructive?*
- *Could her behaviour affect the team? How?*
- *Would you seek advice or support from anyone else regarding the situation?*

what will be recorded. For example, a staff member may raise a personal matter in supervision that could affect their work practice. You will need to agree with them how this will be recorded and who will have access to this information.

VIEWPOINT

- Why do you need to record personal issues that may affect practice?

Supervision notes need to be clearly written, with sufficient detail so that they can be read and understood at a later date. However, it is important to remember that this is the staff member's one-to-one time with you as their manager, and it can be off-putting if you never look at them as you are constantly writing. You will also miss important cues from non-verbal communication, such as facial

ACTIVITY

Games people play

Look at the following examples of games people can play during supervision. These can be played by supervisees and supervisors. Consider these questions.

- Have you ever encountered any of these behaviours?
- What strategies do you have to deal with them?
- Which behaviour types trigger different responses from you?
- Have you played any of these games in supervision, either as supervisee or supervisor? If so, what were the circumstances that led to this behaviour?
- **Be Nice To Me Because I Am Nice To You** The supervisor is flattered by the compliments of the supervisee on what a good supervisor they are, and then finds it difficult to deal with issues around work and appropriate boundaries.
- **Treat Me, Don't Beat Me** The supervisee involves the supervisor in discussion of their personal problems, changing the relationship to one of worker–client and avoiding work issues.
- **Evaluation Is Not For Friends** The supervisory relationship becomes a social relationship and this is used to make it more difficult to maintain the professional component in supervision sessions.
- **You Think You've Got Problems** The supervisor listens with increasing lack of interest to the supervisee's difficulties instead of focusing on their workload/issues, which they view as much more significant. Supervisor changes the relationship to one in which s/he elicits sympathy and support from the supervisee.

- **If You Knew Dostoyevsky Like I Know Dostoyevsky** The supervisee mentions the fact that the parent's behaviour reminds them of Raskolnikov in *Crime and Punishment* which is, after all, somewhat different in etiology from the pathology that plagued Prince Myshkin in *The Idiot*. Such a ploy undermines the supervisor's position because he or she may not have heard of the book and may feel unable to admit to being so ignorant.
- **I Have a Little List** The supervisee chooses a list of topics they know interest their supervisor and the session becomes a one-way flow of lectures on those topics by the supervisor, prompted by the odd question.
- **Heading Them Off At The Pass** The supervisee opens the session by freely admitting their mistakes, confessing they should have done better, done it sooner, done it as the supervisor suggested. The supervisor feels forced to offer sympathy and support and the problem is not properly discussed.
- **Little Me** The supervisee looks to the knowledgeable, competent supervisor for a detailed description of how to proceed by asking questions such as 'What would you do next?'
- **I Did It Like You Told Me** The supervisee does exactly what the supervisor suggests in spiteful obedience and, when it fails (as it is bound to do!), blames the supervisor.
- **It's All So Confusing** The supervisee cites the views of others, saying they find it difficult when they get conflicting advice. The supervisor ends up defending his/her view against an unknown competitor, whose views may not be accurately put anyway.
- **I've Tried That Before** The supervisee fends off ideas for change by citing examples from the past where the action, currently being suggested, has failed.
- **When You've Been Around As Long As I Have** Often used by an older supervisee to a new supervisor. The supervisee sighs and refers to his/her vast experience as a reason for not implementing change or taking a specific action.

(Adapted from Kadushin, 1968; Hawthorne, 1975)

expressions and body language. As with many areas of your work, practice helps to develop the skills required to be an effective manager, including note taking.

Supervision notes should include specific examples of practice, identifying positive aspects of the staff members work and areas to be developed. They should also include records of key areas of discussion, and record decisions made and action points agreed (including identification of the individual responsible for each action point and the time frame for completion).

Remember that appraisals should never have surprises: all issues to do with a staff member's practice should be raised during their regular supervision sessions throughout the year. In this way, good supervision notes can be used to inform the appraisal process.

Supervision notes should be stored on the staff member's file, with a copy given to them for their records – this is especially important if there are action points for them to complete before you next meet. You would have informed them, as part of the agreement process, of who else will have access to their files: for example, your line manager and Human Resources. Other than to identified people, it is important that supervision notes remain confidential. All relationships are built on trust; the one between and manager and their staff team is no different. If the content of a supervision session becomes 'common knowledge' because supervision notes have been left in a place where other staff members can read them, working relationships will be seriously damaged. Files should be kept in secure cabinets, to which only the managers of the setting have access. While all staff have a right to access to their files, they would need to request to see them. This is line with the Data Protection Act 1998 which states 'an individual has right of access to all information held about them'.

Giving feedback

It is good practice to give feedback to supervisees on a regular basis. Wherever possible, first ask the person to assess their own performance. It is good practice to ask for the supervisee's views on how well the work/task was done before giving your own views – this will help you gauge how well the supervisee understands what is expected of them in terms of role and level of performance, and will develop the supervisee's self-evaluation skills.

Individuals are often quite critical of their work and tend not to take sufficient credit for the things they have actually done well. Feedback sessions provide you with an opportunity to give praise and positive affirmation. On the other hand, if there is an issue about the supervisee's standard of performance, it is better to try to get them to identify this for themselves. If they are unable to identify areas of difficulty themselves, the role of the supervisor here will involve challenging the supervisee's perceptions. It is better to give developmental feedback on one or two areas of difficulty at a time, and not overload or demoralise the supervisee with more.

It is essential to keep in mind that individuals perceive and receive criticism differently, and to be sensitive to this, to ensure that individuals do not suffer loss of self-esteem or feel demotivated in their role. It is useful to have an explicit discussion about how individuals prefer to receive feedback at the beginning of the supervisory relationship. Then, remember the feedback 'sandwich'!

First, give positive feedback: what was done well, and why it was right or good.

Be:

- clear – use language appropriate to the person receiving the feedback. Avoid ambiguous terms such as 'that was *quite good*'.
- specific – give particular examples about what was done well, rather than general comments, such as 'That was good' or 'You did well'
- personal – make the feedback personal, from you to the supervisee. Use their name and 'I' and 'You' in your feedback: for example, 'I think that your contribution to the review was clear and helpful.'

Next, give developmental feedback: it would have been 'even better if ... '

Be:

- specific – (as above)
- constructive – explain what changes are required, and check the supervisee understands
- kind – focus on the performance or behaviour, not the person
- personal – own your feedback by using 'I' statements.

Finally, always end with something positive and encouraging, but still be truthful (do not undermine any developmental feedback you have given).

CASE STUDY

Giving feedback

You have carried out an observation of a member of staff. The member of staff was supporting a service user for whom they are keyworker, to complete the 'My Plan' booklet for the service user's review in two weeks' time. The service user has learning difficulties and has some difficulties communicating verbally.

The staff member sat in the sitting room with the service user and had a friendly, warm approach. She sat alongside the service user so that they could see the booklet. She explained what they were doing and why they were doing it. You observed that she didn't check the service user had understood this. She then went through the booklet and asked questions. As the service user attempted to answer a number of questions, the staff member interrupted and answered for him.

- *Which areas of practice would you identify as good, and which areas need ongoing development?*
- *How would you give feedback to the member of staff? What would you say? When and where would you say it?*
- *How will you support the member of staff to develop in the areas identified?*

REFLECT

Reflect on a situation where you have given a supervisee or colleague feedback on a piece of work where you identified some areas for development. Reflect on how closely your feedback followed the principles of the 'feedback sandwich'.

- Could you have followed the principles more closely? If so, how?
- What might you have done or said differently?

Appraisals

Staff appraisals form part of the annual cycle of staff management, and should serve to review the previous year and set goals for the forthcoming year. Organisations use appraisal for a variety of purposes, including:

- **reviewing individual performance and achievements** against goals agreed from the previous appraisal
- **reviewing and planning ongoing professional development** Professional development and learning needs will have been discussed on a regular basis in supervision. However, the appraisal should draw together the discussion and learning activities from the previous year and consider the year to come. This may include identifying development activities to assist the appraisee to fulfil new roles and responsibilities. These might be identified in anticipation of changes resulting from new legal or policy requirements, or through review of the job description (see below). It is also a useful opportunity to consider any qualification requirements attached to the role (such as NVQ/SVQs) or professional development requirements specified by regulatory bodies such as the General Social Care Council (GSCC) or Health Professions Council (HPC). The following

section, *Identifying appropriate learning activities*, will be useful for this area of the appraisal.
- **Providing feedback** – Appraisal provides an opportunity to provide feedback to the appraisee regarding their performance over the previous year, but also for the appraisee to provide feedback about the organisation and service delivery. This feedback is important in maintaining a 'learning culture', and could be significant in maintaining motivation and in identifying issues related to recruitment and retention of staff.
- **Reviewing the job description** – This involves considering whether the individual's job description accurately reflects their current role and responsibilities. Any changes made to the job description as a result of this discussion should be agreed by both the appraiser and appraisee, and be in line with the organisation's human resource procedures.
- **Reward** – The use of appraisals in making decisions around reward, such as performance-related pay, is contentious; some of the literature advises that separate systems should be in place to deal with pay and promotion (Livy, 1988; SCIE, 2005). However, others argue that it is difficult and indeed naive to believe that it is possible to separate performance review and reward review (Randall, 1989; Hunt, 1992). Each organisation needs to make the purpose of its appraisal clear to its employees. It is clear, however, that disciplinary issues should always remain separate from appraisal processes.

Effective appraisals

Here are some of the key elements in ensuring appraisals are effective.

- The appraisal process should be a positive process and experience. Even where the appraisee's performance has been below the required level, the focus should remain on setting goals and identifying strategies to improve that performance over the following year, rather than criticising performance during the previous year.
- There should be no surprises for the appraisee, as any issues should have been dealt with as they arose, and through supervision. The appraisee should have received regular feedback, both positive and developmental through the previous year.
- Feedback regarding performance should follow the 'feedback sandwich' model. Remember to give specific feedback, providing examples to illustrate your points.
- The appraiser should avoid being defensive when receiving feedback about themselves or the organisation. Focus on listening actively and showing the appraisee you appreciate their feedback. You may wish to seek clarification particularly if the feedback is very general, but do not try to defend yourself or the organisation – just listen.

Planning appraisals

It is vital for both the appraiser and appraisee to prepare for the appraisal – organisations often offer training on appraisals. The appraisal date should be booked well in advance of the meeting, with the appraisee given adequate time to reflect on their performance and future goals, and to prepare any notes, such as the self-assessment below.

The appraisal

As in supervision, ensure there is adequate time for the appraisal, an appropriate (and definitely private) environment and that there will be no interruptions.

Beginning the appraisal

Welcome the appraisee and check they are comfortable. Reiterate the purpose of the appraisal and the roles of both parties, emphasising it is a collaborative event. Clarify the main areas and format the appraisal will cover.

Main discussion

This part of the appraisal follows the agreed format, often directed by the format provided by the organisation. When discussing the appraisee's

Planning

- What is the purpose of the review?
- Have I given adequate notice of the date?
- What information do I have about the apparaisee's performance during the review period?
- What information has the appraisee provided?

Preparation

- Have I read all the information?
- What are the issues I want to cover in the interview?
- What are the issues the appraisee will wish to cover?
- Do I need to consult with other staff with whom the appraisee works closely to determine potential objectives and performance indicators?
- Do I need to consult with other staff to agree standards?
- Have I thought of a provisional plan for the review?
- Has the appraisal been arranged in a suitable room and all possible interruptions diverted?

Appraisal

- How am I going to put the individual at ease?
- How do I ensure the appraisee understands the purpose of the appraisal scheme?
- Am I going to agree to a plan for the review with them?
- How am I going to incorporate their responses?
- What are my listening difficulties and how can I overcome them?
- How do I ensure that the appraisee contributes fully to the discussion?
- How will I use open, closed, probing and checking questions?
- What are my prejudices and how can I set these aside?
- How will I take notes of the discussion without distracting the appraisee
- How will I deal with giving and receiving criticism?
- Should I summarise at different stages of the interview?
- How can I ensure the interview will end with a clear understanding and agreement on the issues and agreed plan?

Follow-up action

- How can I ensure all the documentation is completed within the timescale?
- How will I ensure that the issues raised at the interview are actioned?

(SCIE, 2005)

performance over the previous year, ensure that the appraisee begins the discussion. This is important in helping the appraisee to reflect on their own performance and practice, and for you, as an appraiser, to know how they think they are performing.

Conclusion

Summarise the main points from your discussions, and confirm the goals for performance and professional development for the next year. Remember to finish on a positive note. It is useful to schedule a review point, usually mid-way through the year, to discuss progress, any issues arising and adapt plans or goals as required.

Writing goals

Take care not to identify too many goals, which may feel overwhelming and demotivating to the appraisee; 4–6 goals is ideal. Also remember that goals should be SMART:

- **S**pecific – what exactly needs to be achieved? Try and be as precise as possible.

- **M**easurable – how will you know that the goal has been achieved? If you cannot answer this, the goal may need to be re-written.
- **A**chievable – the goal may stretch the appraisee a little, and this is likely to help them feel motivated; however, if the goal feels totally unattainable – for example, completing an NVQ/SVQ in three months – it is likely to demotivate them.
- **R**elevant – the goals should be relevant to the appraisee's role and responsibilities, and to the organisation's overall goals and business plan.
- **T**ime-bound – the timescale for achieving the goals should be agreed and clearly recorded.

Recording appraisals

Organisations normally produce a standard form to record appraisals. Figure 4.7 is an example of an appraisal record.

SCIE (2005) gives a useful list of questions for managers to consider when planning and preparing for an appraisal.

ACTIVITY

Look at your own learning

Before considering how to facilitate the learning of others, it is helpful to consider how you learn. Answer each of these questions.

- Think of a recent positive experience where you learned something new. What aspects of the experience made it positive?
- How do you approach new learning? Do you have particular strategies or orders in which you do things? Do you always approach learning in the same way?
- What do you find easy to learn? Why?
- What do you find difficult to learn? Why?
- How will this inform your work with others?

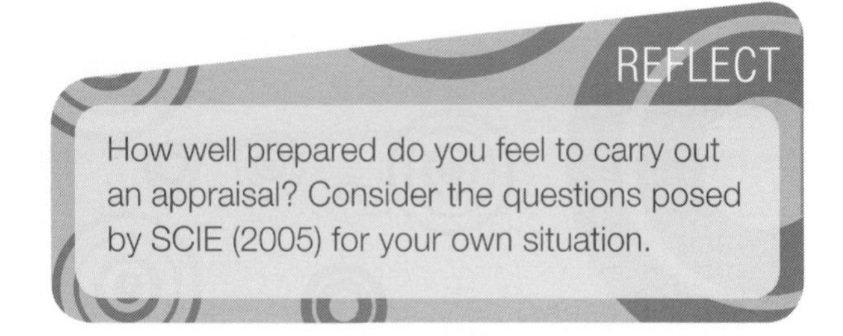

Learning activities

Learning needs should be identified on an ongoing basis, from the point of recruitment right through the individual's employment with the organisation. Supervision and appraisal are the most significant opportunities to identify learning needs and plan methods to meet them. It is important to think in terms of 'learning activities' rather than just 'training': training is just one of a range of methods to address learning needs.

Before considering learning preferences and activities, it is useful to define what we mean by learning. Burns (1995) provided a useful definition of learning as 'a relatively permanent change in behaviour, with behaviour including both observable activity and internal processes such as thinking, attitudes and emotions'.

Learning styles

The activity on page 67 may have highlighted that you have particular preferences when it comes to learning. The work of Kolb (1984) and Honey and Mumford (1986) explains such preferences in more depth.

You met Kolb's Learning Cycle in Chapter 1 (page 3). As you may remember, Kolb explains learning as a continuous process or cycle, with four identifiable stages.

Appraisal and Development Record

Personal details

	Appraisee	**Appraiser**
Name		
Job Title		

Date of appraisal	**Date of last appraisal**	**Date of interim review**	**Date of next appraisal**

Self-assessment *(To be completed by appraisee before the meeting)*

1. **Current goals** (from previous appraisal)
   a)
   b)
   c)
   d)
   e)

2. **Summary of overall performance**

   Main achievements

   Main challenges

   Other issues you would like to raise

Appraisal

1. **Performance against goals**
2. **Overall performance**
3. **Goals for the next year**

Goal	**Success criteria**

4. **Learning and development plan for the next year**

Signatures

Appraisee: Date:

Appraisor: Date:

Appraisor's manager Date:

Figure 4.6 A sample appraisal and development record

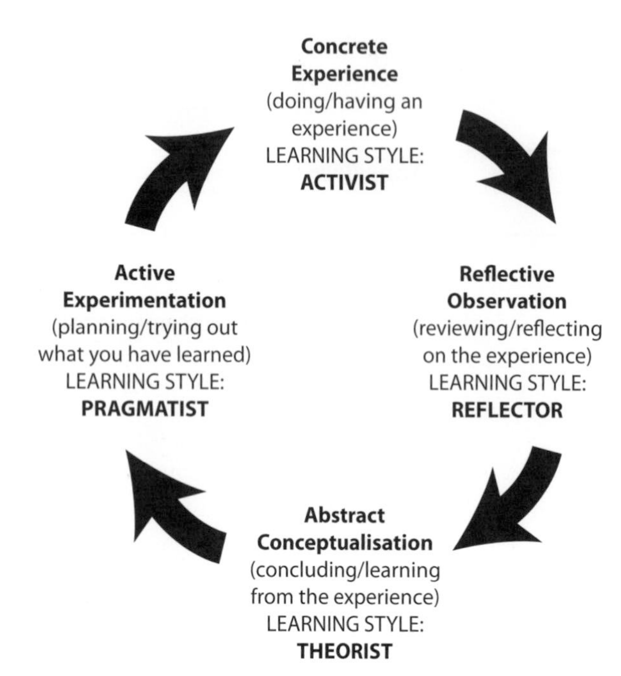

Figure 4.7 How Kolb's Learning Cycle relates to learning styles

Concrete experience refers to actually having the experience, or the 'doing'.

Reflective observation refers to the process of reflecting on the experience: for example, considering what went well, and did not go well or as expected.

Abstract conceptualisation refers to the process of drawing in our knowledge base to help us understand why things happened as they did. For example, how does theory or research help our understanding of this situation?

Active experimentation refers to the stage at which we plan the next course of action. This might involve considering options, trying things out and making decisions about the next stage before we move on to 'concrete experience' again, that is, 'doing'.

Individuals do not go through the learning cycle in clear smooth stages, or for every aspect of learning they undertake. They may feel more comfortable at certain points of the cycle, and therefore tend to focus more on that stage. There are critics of Kolb, such as Holman et al., (1997) and Kayes (2002), as well as other theories of learning, such as reinforcement theory (Laird, 1985), which you may wish to review. However, Kolb's cycle does provide a helpful model for understanding adult learning.

From the learning cycle, Kolb developed the idea of a number of learning styles. Honey and Mumford (1986) built on Kolb's work to create learning styles questionnaires and exercises to enable individual to identify their preferred learning style. Honey and Mumford refer to four learning styles.

- **Activists** enjoy the experience itself. They want to be 'doing' and having new experiences and challenges. You may think of them as people who like to 'dive straight into' things.
- **Reflectors** spend time and effort reflecting. They are likely to step back and observe, and consider past events and different perspectives before 'doing'. They are likely to listen to others first before contributing to a discussion.
- **Theorists** are good at making connections and abstracting ideas from experience. They focus on the underpinning models, principles and theories to inform their thinking. They may seem more objective, but also cautious.
- **Pragmatists** enjoy the planning stage, and like to come up with new ideas and ways forward. They are keen to test out their ideas and see if they work, so may become impatient with the reflectors and theorists.

REFLECT

- Can you identify which learning style fits you most closely?
- If you have completed a learning styles questionnaire, are the results as you expected? How does this inform your thinking and practice?
- See the resources listed at the end of the chapter for details of the learning styles questionnaire.

Of course, you may see elements of more than one style in your own learning, or see that, in different situations, you draw on different styles more fully. In addition, experience may assist you in being more flexible in drawing on the range of learning styles – and one of the skills of management is being adaptable to enable staff to develop. However, Honey and Mumford (1986) suggest that individuals tend to have one preferred style, which may be the style you revert back to when under pressure.

Having considered the learning cycle and learning styles, now consider different types of learning activities in the Activity panel below.

It is important not just to plan learning activities but also to ensure that individuals reflect on their learning following the activity. Supervision is the ideal environment for this reflection, and will also help the supervisee to evaluate whether learning has taken place, and whether the activity may be useful for other members of staff. However, remember that learning may not be displayed in observable behaviour until some time after the 'learning event' has taken place. Here appraisals (and mid-year appraisal reviews) will be useful in confirming whether learning took place, and whether learning continues to be applied in practice.

Exit interviews

Exit interviews are generally accepted as good employment practice. Although organisations aim to retain staff, there is inevitably some natural turnover within staff groups – and as previously mentioned, in some areas of social care, turnover of staff is high. Therefore, it is important not to neglect the end stage of a person's employment with you, as there may be much to learn from it.

The aim of exit interviews is normally to:

- identify the reasons why a member of staff is leaving the organisation
- spot any trends or patterns in reasons for staff leaving, and thus help to address any areas in the organisation which need improvement.

ACTIVITY

Learning activities

Consider each of the types of activity listed below and identify which types of learning needs they might be most appropriate in meeting.

- Will the activity help the development of skills, knowledge or both?
- Which learning styles would the activity most suit?

Activity	Appropriate for ...
Shadowing others	
Reading	
Simulation, e.g. role play	
Joint working	
Coaching or mentoring from others	
Project work, e.g. a specific, short term piece of work or research	
New task	
Seminars or short courses	
Visits to other organisations/ establishments	
Reflective writing	
Qualifications	
Secondment	
Other: specify any other activity	

Remember that staff have a choice about whether they wish to have an exit interview; they cannot be compelled to do so.

The exit interview

The exit interview should provide the member of staff with an opportunity to express their views in a confidential arena, and the interviewer with an opportunity to listen to real concerns. Therefore, it is vital that exit interviews are seen as independent and confidential. This also reinforces to other employees that the organisation is willing to learn and improve: a sign of a 'learning organisation'. Interviews should take place in a private room, free from interruptions, and with enough time allowed for the member of staff to discuss their reasons for leaving and provide feedback on their experience of the organisation. The purpose of the interview and boundaries of confidentiality should be made clear at the beginning of the interview.

Organisations usually have a set of standard questions to be used in exit interviews.

Here are some examples.

- What is your main reason for leaving your current job?
- Are there any other reasons?
- How much influence did each of the reasons have on your decision to leave?
- Did your current job description accurately reflect your day-to-day role and responsibilities?
- Could your skills and knowledge have been used more effectively by the organisation? If so, how?
- What have you enjoyed about working with the organisation?
- What has been difficult or frustrating about working with the organisation?
- What attracted you to your new job?

Exit interviews should be recorded fully so that common trends and responses can be analysed. Analysis of exit interview records should take place regularly and consistently – perhaps once or twice a year – so that the organisation can identify any issues or problems and put appropriate action plans into place.

ACTIVITY

Dealing with exit interview feedback

You receive feedback from the senior management group of your organisation that exit interview analysis has shown the following main reasons for staff leaving their posts:

- lack of opportunities for development
- lack of flexible working
- the presence of a 'blame culture'.

In your role as manager, what strategies could you put into place to address these issues within your staff team?

Figure 4.8 An exit interview is a perfect opportunity to collect opinions and information on your organisation

References

Beverly, A. and Coleman, A. (2005) *Induction to Work-based Learning and Assessment Workbook for Assessors* (2nd edition), Leeds: Skills for Care

Brown, A. and Bourne, I. (1996) *The Social Work Supervisor*, Buckingham: OU Press

Burns, R. (1995) *The adult learner at work*, Sydney: Business and Professional Publishing

Collins, J. (2006) *Good to Great and the Social Sectors*, Colchester: Random House Business Books

CSCI (2006b) *In Focus: Quality Issues in Social Care, Issue 4: Safe and sound? Checking the suitability of new care staff in regulated social care services*, London: CSCI

DOH (2003a) *National Minimum Standards: Care Homes for Adults*, London: TSO

DOH (2003b) *National Minimum Standards: Care Homes for Older People*, London: TSO

DOH (2007c) *Social Care Workforce Initiative Newsletter* (June 2007) www.dh.gov.uk/en/SocialCare/Aboutthedirectorate/Researchanddevelopment/index.htm (accessed 15.09.08)

DOH (2008), *Social Care Workforce*, www.dh.gov.uk/en/SocialCare/workforce/index.htm (accessed 15.09.08)

DOH and DFES (2006) *Options for Excellence: Building the Social Care Workforce of the Future*, London: DH/DFES

Hawthorne, L. (1975) 'Games Supervisors Play', *Social Work*, 20(3):179–183

Holman, D., Pavlica, K. and Thorpe, R. (1997) 'Rethinking Kolb's theory of experiential learning in management education: the construction of social constructionism and activity theory', *Management Learning* 28(2):135–148

Honey, P. and Mumford, A. (1986) *Manual of Learning Styles*, Maidenhead: Peter Honey

Hunt, J.W. (1992) *Managing People at Work* (3rd Edition), Berkshire: McGraw-Hill

Kadushin, A. (1968) 'Games people play in supervision', *Social Work*, 73:127–136

Kayes, D.C. (2002) 'Experiential Learning and Its Critics: Preserving the Role of Experience in Management Learning and Education', *Academy of Management Learning and Education*, 1(2): 137–149

Laird, D. (1985) *Approaches to training and development*, Reading, Mass: Addison-Wesley

Livy, B. (1988) *Corporate Personnel Management*, London: Pitman

London Councils (no date) *Workforce Analysis and Planning guide*, www.londoncouncils.gov.uk/improvement/hrip/workforceanalysis/default.htm (accessed 27.09.08)

National Institute for Social Work (1996) *Managing Social Work: Briefing no.14*, London: NISW

Randall, G. (1989) 'Employee Appraisal'. In Sisson, K. (ed) *Personnel Management in Britain*, Oxford: Basil Blackwell

SCIE (2004) *Learning Organisations: A self-assessment resource pack*, www.scie.org.uk/publications/learningorgs/index.asp (accessed 27.09.08)

SCIE (2005) *Performance Appraisal in a Nutshell*, www.scie-peoplemanagement.org.uk/resource/docPreview.asp?surround=true&lang=1&docID=128 (accessed 27.09.08)

SCIE (2006) *Common Induction Standards: Guidance for those responsible for workers in an induction period*, Leeds: Skills for Care

SCIE (2007) *Workforce Planning: A mini guide*, www.scie-peoplemanagement.org.uk/resource/docPreview.asp?surround=true&lang=1&docID=137 (accessed 27.09.08)

Scottish Executive (2005b) *National Strategy for the Development of the Social Service Workforce in Scotland*, Edinburgh: Scottish Executive

Senge, P.M. (1990) *The Fifth Discipline: The Art and Practice of the Learning Organization*, New York: Doubleday Currency

Skills for Care and CWDC (2007) *Providing Effective Supervision*, Leeds: Skills for Care/CWDC

Thompson, N. (2006) *Promoting Workplace Learning*, BASW: Policy Press

Topss (2004) *Workforce Planning Toolkit* (2nd edition), Leeds: Topss

Useful reading and websites

SCIE has produced a useful self-assessment resource pack on learning organisations (2004) *Learning Organisation Self-Assessment Resource Pack*

Honey and Mumford, *Learning Styles*, http://www.peterhoney.com/

http://www.skillsforcare.org.uk/view.asp?id=58 (accessed 23.11.07)
Collins, J. (2006) *Good to Great and the Social Sectors*, Colchester: Random House Business Books
DOH (2008) *Putting People First - working to make it happen: adult social care workforce strategy – interim statement*, London: DOH
SCIE (2004) *Learning Organisations: A self –assessment resource pack.* www.scie.org.uk/publications/learningorgs/index.asp
SCIE (2006) *Common Induction Standards: Guidance for those responsible for workers in an induction period*, Leeds: Skills for Care
SCIE (2008) *People Management*, www.scie.org.uk/workforce/peoplemanagement.asp

CHAPTER 5

Common challenges in management

'Good people management ultimately results in an improved experience for service users, but can be a real challenge to achieve in the social care sector.'
(SCIE, 2008:1)

Introduction

Chapter 4 gave you an overview of the key areas of workforce planning and development. Chapter 5 examines some of the more challenging areas of performance management in more depth. The chapter covers:

- team development
- change management
- conflict management
- motivation
- delegation
- managing poor performance
- disciplinary and capability processes.

Team development

Much of the work undertaken in social care is done in teams. To achieve the aims and objectives of the organisation or service effectively and efficiently, it is vital that teams work well together. Team building and development is not a single event, but an ongoing process that takes place within the team over a long period of time.

ACTIVITY

Successful teams

Think of a well-known, successful team and list the main features or factors which you think make the team successful. Working with study colleagues or colleagues in your team compare the factors you have identified with those identified by colleagues thinking of other teams.

- Are there factors which the teams seem to have in common, and if so which?
- Are any of these factors/features present in your current team?
- How do you think these factors/features could be developed within your team?

To manage teams effectively, it is helpful to consider how teams develop, and the contribution individuals make to team working.

Tuckman's Stages of Team Development

One of the best-known theories of team development was developed by Bruce Tuckman (1965), who identified a number of stages that groups go through during their existence. Teams may not be conscious of these processes, but an awareness of the stages may help teams to deal with the more challenging aspects of team working.

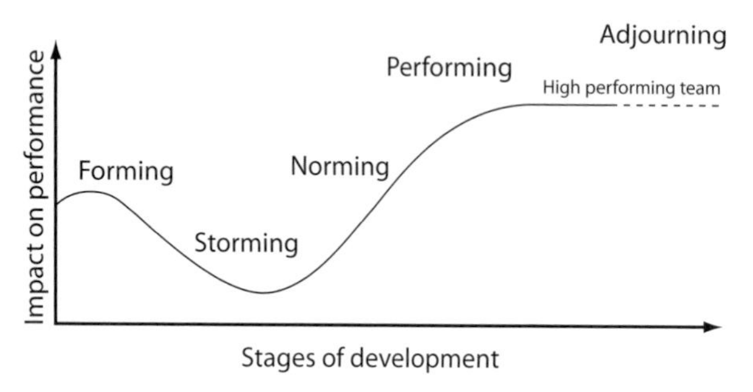

Figure 5.1 Tuckman's Stages of Team Development

Stage 1: Forming
This is when individuals are exploring what other team members expect of them, and what the norms and rules of the group are. These norms and rules may be unspoken, making this stage more difficult, and possibly anxiety-provoking, for individuals. Individuals are also trying to establish their own identity within the group at this stage.

Stage 2: Storming
At this stage, conflict occurs while the team members try to establish themselves within the group and explore roles and relationships. Sub-groups may form, challenge each other and seek power within the team. This can be a difficult and challenging stage for managers.

Stage 3: Norming
The team begins to develop 'norms': behaviours, 'rules' or beliefs that are accepted by the group, but are not formalised. The group becomes more cohesive and cooperative at this stage. Conflicts minimise, and the team becomes more supportive, with an increasing sense of team identity.

Stage 4: Performing
The team is working effectively together, and is focused on tasks and overall aims. Where there are disagreements, they are dealt with openly and constructively. The team manager will be able to delegate tasks more easily at this stage.

Stage 5: Adjourning
Tuckman and Jensen (1977) added this stage to his model to reflect the point at which a team may have come to an end of its work/project, and is disbanding. This can be a difficult and upsetting stage for individuals if the team has bonded well. It should also be seen as a stage where the success of the team can be recognised and celebrated.

Teams will not move through these stages at the same pace, and may move back and forth between the stages. If individuals leave and join the team, this can also impact on team development, resulting in what is sometimes termed 're-forming'.

ACTIVITY

What stage is your team at?

Look at Tuckman's Stages of Team Development.

- Identify the stage your team is at.
- Why do you think it is at this stage?
- Read the rest of this section on team development and then identify strategies you can use to assist your team to move to Stage 4 (performing), if they are not there already.

Belbin: Team Roles and Individual Functions in a Team

Dr Meredith Belbin (1981; 1993) identified nine different roles people play in teams. Belbin argued that individuals make different contributions to the team (and team task) according to their personal characteristics.Teams work best when there is a balance of roles represented within the team, and when team members know their own roles, work to their strengths and actively manage weaknesses.

It is important to remember that the roles individuals adopt will depend to some extent on the situation they are in, their relationship with others in the team and the type of work being undertaken. However, reflecting on the individuals within your team and the roles they play may help you to understand the functioning of the team better, and help you to get the most from the team.

Team-building events and activities

Although team development is an ongoing process, it is also important to set aside specific times for team-building activity. This might be whole- or half-day events, meetings, team lunches or social events. When planning team-building events and activities, it is important to identify aims/objectives that match

Overarching focus	Belbin roles	Description: strengths and weaknesses
Doing/acting	Implementer	Well organised and methodical. Takes basic ideas and makes them work in practice. Can be slow and inflexible.
	Shaper	Is active and dynamic, and challenges others to move forward. Can be insensitive.
	Completer/Finisher	Reliable and detailed, sees things through to the end, and ensures everything works well. Can worry too much and finds it difficult to trust and delegate to others.
Thinking/problem solving	Plant	Creative, and solves difficult problems with new ideas. Can be poor at communicating with others, and may ignore details.
	Monitor/Evaluator	Is a strategic thinker, and considers all options thoughtfully. May lack drive and can have difficulty inspiring others.
	Specialist	Has expert knowledge and skills in key areas; is focused and dedicated. Tends to focus on their specialist areas only.
People/feelings	Coordinator	Respected, confident leader; helps to maintain focus on goals. May be seen as manipulative and delegating too much work to others.
	Team worker	A good listener and diplomatic; works to resolve problems and maintain relationships. Can be indecisive.
	Resource/investigator	A good networker, with an optimistic outlook; explores new ideas and opportunities. Can be too optimistic, and loses momentum after initial interest has waned.

Figure 5.2 Belbin's nine team roles

ACTIVITY

Team roles

Consider these questions.

- Which team roles are represented in and which are missing from your team?
- Is there a role(s) that many of the team members share?
- Are any roles particularly important to the function of the team or nature of work undertaken by team?
- Which strengths are present and which are missing from the team?
- What impact might this have on the work undertaken by the team?
- Can you identify any strategies you could adopt to develop strengths and minimise weaknesses?

the needs and issues within the team, and will make a long-term contribution to better team working. It is also important to consider the most appropriate activities to address the aims.

Figure 5.3 Team building can take many forms

Aims might be to:

- generate and share ideas
- sustain a feeling of team identity
- provide ongoing support
- increase confidence
- improve working relationships
- provide information and updates
- plan and discuss changes
- review critical incidents
- address communication issues
- reward the team for good work and celebrate success.

CASE STUDY

Team-building scenarios

Consider these two scenarios, and answer the questions below.

Scenario one *You work in a residential care home for people with learning difficulties. The home has recently been inspected and criticised for failing to take a person-centred approach to care planning. All teams within the organisation must produce an action plan to improve their performance in this area.*

Scenario two *You work in a residential care home that provides care and support to adults recovering from drug and/or alcohol addiction. Until six months ago, you had a very stable and experienced team with close working relationships. The team had worked together for over two years until two members left the organisation within a few weeks of each other. Two new members of staff were recruited to the team fairly quickly, but previous team members obviously miss their former colleagues and continue to maintain their friendships outside of work. The new team members seem to be struggling to integrate fully into the team.*

1. *How might team members be feeling in each case?*
2. *How might Tuckman's and Belbin's theories inform your understanding of the situations?*
3. *Choose one of the scenarios and plan a team event and activities to meet the needs of the team. Present this to a group of colleagues, justifying your plan and choice of activities.*

When planning a team-building event, you will also need to consider:

- whether the event and activities will achieve the aims/change you want
- the time, budget and venue required
- how team members will be involved in planning and delivering the event
- how ground rules will be negotiated/agreed
- whether individuals might feel excluded: for example, will all individuals feel able to take part in certain social activities?
- how you will evaluate the event
- what the indicators or measures of success are
- how you will ensure that learning from the day will be transferred back to the workplace
- a contingency plan in case something does goes wrong and aims are not achieved.

Team effectiveness

All of the activities in this section relating to teams will help you to reflect on and improve the effectiveness of the teams to which you belong. The University of Victoria, Canada (2004) has produced a helpful diagram to represent team effectiveness (Figure 5.4). You might find it useful to consider the strengths and weaknesses of your team in relation to this model, and identify areas for ongoing improvement.

NVQ/SVQ

Work products you may generate, and either include in your portfolio or show your assessor to demonstrate your skills and knowledge, are:

- **minutes of team meetings**
- **records of team development days, including an outline of the structure of the day and evaluation/feedback from team members.**

Change management

Working in social care means working with constant change. Managing change is something that individuals, teams and organisations can find challenging and anxiety-provoking. Assisting individuals and teams to cope with change is also an area that managers can find particularly challenging.

REFLECT

Think about a significant change you have been involved in (but not as a manager leading change). List some of the positive and negative features of the change.

- What does this tell you about how the change was introduced and managed?
- What lessons can you learn from this in your practice as a manager?

Why change?

Change occurs for a variety of reasons, including:

- in response to new regulations, policies or legislation. This may be local – within your organisation – or national, such as the implementation of a new piece of legislation or policy: for example, Care Standards Act 2002 or *Transforming Social Care* (DOH, 2008d).
- as a result of new research or good practice guidance, which identifies key issues and suggests more effective ways of working: for example, SCIE research briefing 21: *Identification of deafblind sensory impairment in older people* (Roberts, D., et al. 2007).
- a 'disturbance' of a system, such as when part of a building becomes unusable due to flood damage
- as a method of self-preservation: for example, if there is an overspend and savings need to be made

Team Effectiveness Model

Teams can continuously improve their effectiveness by focusing on improving their functioning in five key areas: Goals, Roles, Procedures, Relationships and Leadership:

Goals:	What the team aspires to achieve
Roles:	The part each member plays in achieving the team goals
Procedures:	The methods that help the team conduct its work together
Relationships:	How the team members 'get along' with each other
Leadership:	How the leader supports the team in achieving results.

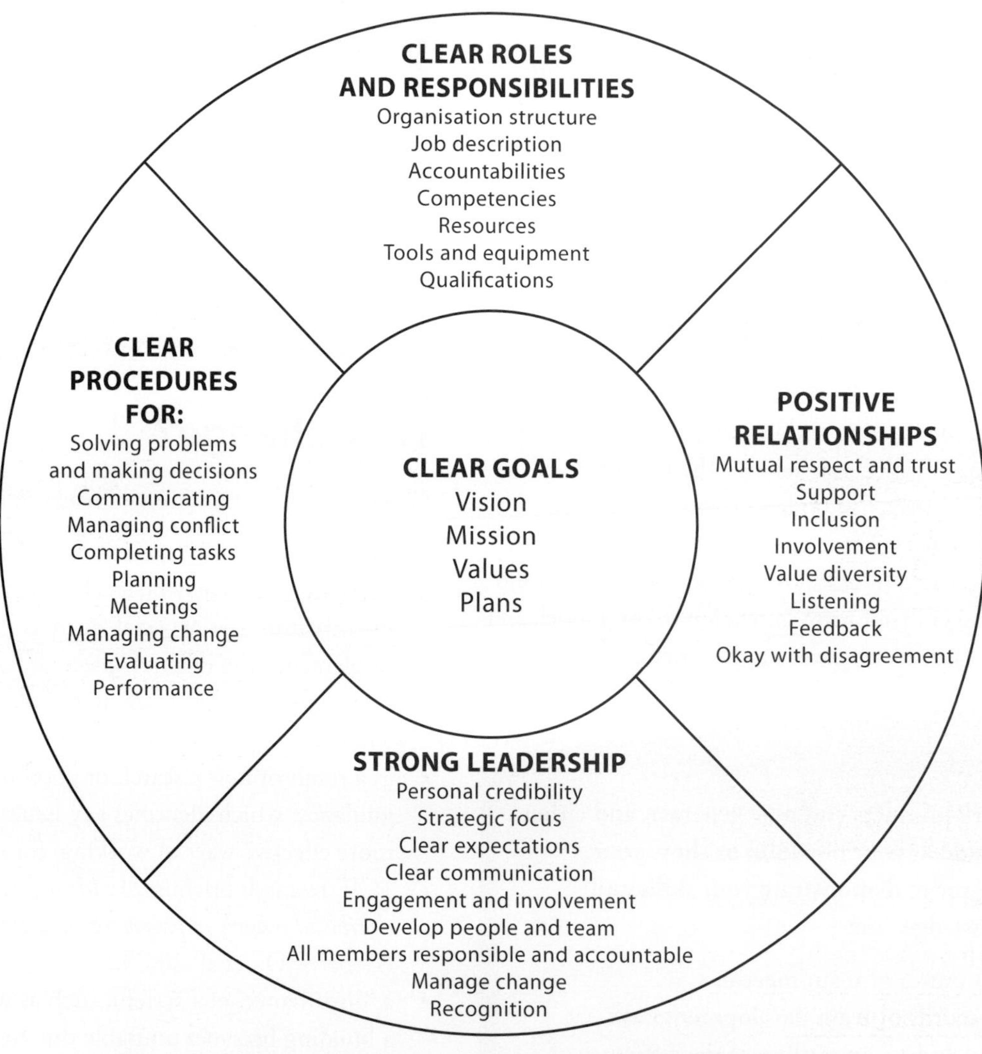

Effective teams are always of and responsible to both their internal and their external environment

Figure 5.4 Model of team effectiveness produced by University of Victoria, Canada

- natural progression and evolution: for example, when staff leave to gain experience of working with a different service-user group
- other external or internal drivers, such as turnover of staff.

Types of change

Change may be planned or unplanned, and can be divided into three main types as shown in Figure 5.6.

Wilfred Krüger developed the idea of the 'Change Management Iceberg' (1996). He believed that, when dealing with change, managers tend to consider a number of 'hard' factors at the tip of the iceberg: cost, quality and time; but, Krüger believed, 'soft' factors such as values, attitudes, emotions and capabilities, which lay below the surface, were much more significant.

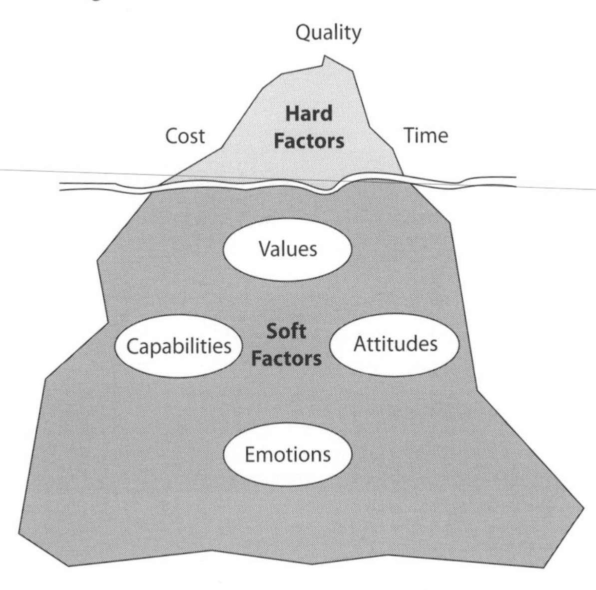

(Adapted from De Witt and Meyer, 2004)

Figure 5.5 The 'Change Management Iceberg'

REFLECT

Why might 'soft' factors be more significant than 'hard' factors in managing change?

What different strategies/approaches might you need to adopt to manage these different types of factors?

ACTIVITY

Team roles and change

- Can you identify where your team members would sit within Krüger's promoters/opponents model (Figure 5.6)?
- Look again at Belbin's team roles (Figure 5.2). Can you identify the specific part that individuals fulfilling each of these roles would play in implementing change? For example, those individuals who welcome change most enthusiastically may not be the best individuals to work consistently and to ensure change is implemented in detail.

As you can see, managing people is a key area in managing change. Individuals will experience change differently, and managers may see a variety of different responses. There are a number of models that may help you to understand how individuals might respond to change.

Krüger (1996) also identified a number of types of individuals who can have a positive or negative effect on the change process.

Type of change	Focus	Example
Incremental	'tweaking' the system or process	adapting a form to collect fuller information
Transitional	restructuring, reorganising	introducing a new shift system
Transformational	new vision, new mission, new values	opening a semi-independent unit

Figure 5.6 The three main types of change

Promoters have a positive attitude towards the change and support it.	**Opponents** have a negative attitude towards the change. Managers need to work with opponents to help change their perception of, and attitude towards, change.
Potential promoters have a positive attitude towards the change but need to be reassured about the benefits of it. Managers will need to encourage these individuals to continue to work towards achieving the change.	**Hidden opponents** have a negative attitude towards the change, although they seem to be supporting it on a superficial level. Managers need to provide clear, explanatory information regarding the change, and to work with these individuals to help change their attitude towards it.

Figure 5.7 Krüger's promoters/opponents model

Change Diagnostic Tool

The Change Diagnostic Tool (Figure 5.8) outlines a number of stages that individuals and teams may move through during the change process. Different behaviours present themselves during each stage. If managers can identify the stage an individual or team is at by observing behaviours, they can adapt their approach to manage individuals/the team more effectively. Managers need to be aware that individuals and teams will not necessarily move through each of these stages, or spend equal amounts of time at each stage, and may move backwards and forwards between stages.

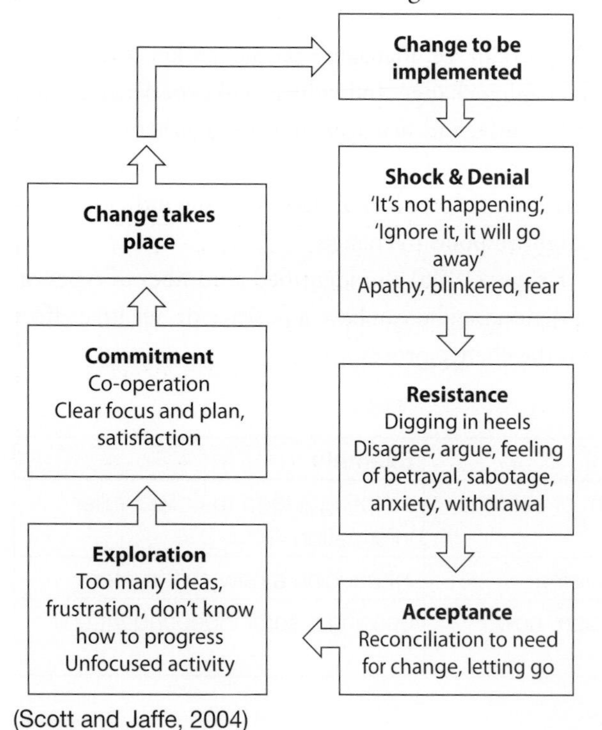

(Scott and Jaffe, 2004)

Figure 5.8 Change Diagnostic Tool

CASE STUDY

Change Diagnostic Tool

Read these statements, which have been made by members of staff in reaction to being told about a change that needs to take place.

- *'Will I be made redundant?'*
- *'Here we go again!'*
- *'I think this is going to work well'.*
- *'How much extra work is this going to be?'*
- *'It's too complicated – how are we supposed to do this?'*

Identify which stage of the change process you believe the member of staff is at and how you might respond and work most effectively with the member of staff.

Planning for change

One of the most effective methods of managing change is to plan for it. Scott and Jaffe (2004) have usefully outlined five stages in managing the change process (see Figure 5.9).

Good practice

Figure 5.10 identifies the key elements of managing change successfully and methods of minimising resistance.

Stage	Overall focus	Manager activity
Aligning	Identify the reason for and purpose of the change, and develop a vision of what the change will look like when it is implemented	• Prepare staff and describe the change as fully as possible • Identify how individuals and the team normally react to change • Assess their readiness for change • Do not plan any additional changes that are not crucial
Planning	Bring team members together to discuss the change and plan a strategy for its implementation	• Make contingency plans • Encourage participation and suggestions • Identify the skills and knowledge that will be needed to implement the change • Plan for the impact of change on performance and productivity • Identify how success will be measured
Designing	Plan new structures and roles	• Create a core group to oversee the change, if appropriate • Draw up new policies and procedures • Continue to remind individuals why change is taking place • Create any new structures needed to support the change
Implementing	Making change happen!	• Try things out, pilot options if appropriate, encourage creativity • Encourage ownership of change • Provide more feedback to individuals and the team than normal • Allow for resistance and provide time for reflection • Monitor the implementation of change
Rewarding	Show recognition that the team has made change happen and work. Celebrate the success!	• Acknowledge the success and results of change in all appropriate fora • Review the learning • Identify and encourage best practices, and share with others • Celebrate publicly

(Adapted from Scott and Jaffe, 2004)

Figure 5.9 Planning for change

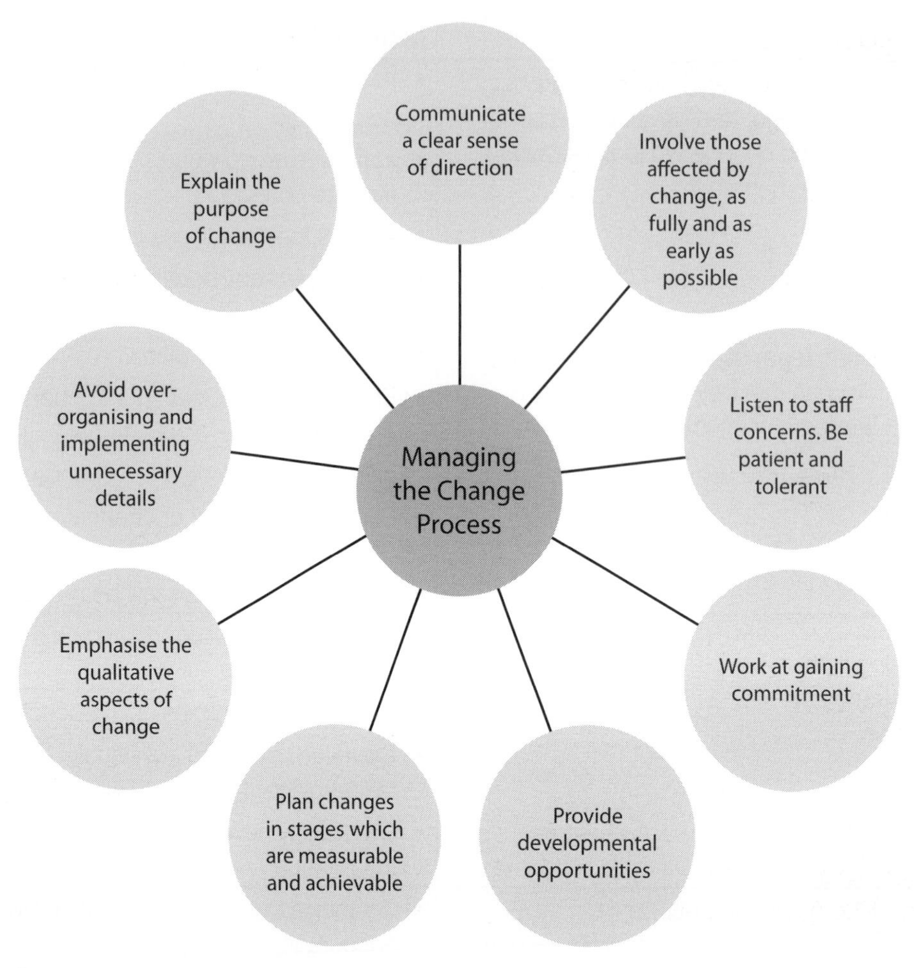

(Adapted from Scragg, T., 2001)

Figure 5.10 Managing the change process

Conflict management

When people think of conflict, it is often seen as something negative and best avoided. However, conflict can be positive and can help to:

- raise and address problems
- encourage the development of creative solutions
- focus attention on key issues, and motivate teams to work on these
- increase participation, where individuals feel that they can have a voice and can influence the way things will develop
- highlight and encourage individuals to recognise and value differences.

Conflict becomes a problem when it begins to:

- adversely affect service users
- lower morale and team identity
- cause increased and ongoing conflict
- cause inappropriate behaviour
- negatively impact on individual and team performance.

A study by CIPD (The Chartered Institute of Personnel and Development) in 2007 identified the most common causes of conflict at work:

- general behaviour and conduct issues
- conflicts over performance
- sickness absence and attendance
- relationships between colleagues
- theft and fraud

ACTIVITY

How do I act in conflicts?

Think of a situation at work where there has been a conflict, and identify which of the five responses below is closest to your reaction.

- Avoid it – you ignore the conflict or pretend it is not occurring.
- Accommodate it – you give in to others, sometimes to the extent that you compromise yourself.
- Competing (challenging it) – you work to get your way, rather than clarifying and addressing the issue.
- Compromising with it – you work to achieve mutual give and take.
- Collaborating with it – you focus on working together, and maintaining working relationships.

Consider these questions too.

- Why did you react in the way you did?
- Was it a conscious or subconscious response? Was it a planned or reactive response?
- Did your response result in the best possible outcome?
- What, if anything, do you think you could have done differently? What learning can you take from this for the future?

- bullying and harassment
- sex discrimination and equal pay issues.

CIPD also reported that performance issues are rated more highly as a frequent cause of conflict among private services and not-for-profit organisations, compared with the other two main sectors and, in particular, public services employers.

Kenneth W. Thomas and Ralph H. Kilmann (1974) identified a number of ways in which people respond to conflict. Figure 5.11 helps you to identify situations when it is most appropriate to respond in each way.

Four key steps

Here are four key steps in conflict resolution.

1. Describe and define the nature of the conflict.
2. Listen actively to all parties involved.
3. Secure the commitment of parties to resolving the conflict by emphasising the benefits of finding a solution or way forward, and the negative consequences of not doing so.
4. Reach an agreement.

Response to conflict	When to use
1. Avoid it	When it simply is not worth the effort to argue. This approach can increase conflict over time and increase your own frustration and dissatisfaction.
2. Accommodate it	Only in situations when the issue is insignificant or when you will be able to adopt another, more effective approach in the near future. This approach can increase conflict over time and increase your own frustration and dissatisfaction.
3. Competing	When you believe strongly that your position is 'correct'.
4. Compromising	When the primary goal is to move on.
5. Collaborating	When the goal is to meet as many current needs as possible by working together and sharing skills, and when the goal is to develop commitment and ownership.

Figure 5.11 Responses to conflict and when to use them

CASE STUDY

Managing conflict

For the past year, you have been managing a cohesive team of individuals who work well together. A new member of staff joins your team, and within a short period of time you notice that team dynamics have changed and there appears to be tension between individuals that was not present previously.

One team member informs you that the new member of staff spends a lot of time gossiping to other staff, particularly criticising your management practice. Another team member informs you that they are feeling very unsettled and are considering leaving because of the disruption and tension caused by the new team member.

- What do you think might be happening here and why?
- What immediate action might you take?
- Would you approach the new team member individually? If so, how and why?
- How will you work with the team as a whole to resolve the situation and restore good team working?

NVQ/SVQ

This section will help to provide evidence for units A5, B4.

Work Products you may generate and either include in your portfolio or show your assessor to demonstrate your skills and knowledge are:

- **records of supervision notes or meetings dealing with issues of conflict.**
- **records of team-building activities and events.**

Motivation

Motivation is a key element of performance. Motivated staff are likely to be more productive and committed, and to have lower levels of absence. Motivation is time- and situation-specific: levels of motivation change depending on each individual's circumstances. Individuals are also motivated by different factors.

Managers play a crucial rule in motivating staff, and theories of motivation can help managers to understand and identify the differences between individuals.

Theories of motivation

Motivational Needs Theory

McClelland (1988) identified three types of motivational need and suggested that each person fits into one of the relevant categories (see Figure 5.12).

Type of motivational need	Characteristics
Achievement-motivated	Desire for excellence, likes doing a good job, wants to advance career, needs feedback
Authority-motivated	Likes to lead, wants prestige and job status, enjoys influencing people and organisation direction
Affiliation-motivated	Likes to be popular and well thought of, enjoys teamwork rather than lone working, likes good relations between team members

Figure 5.12 Types of motivational need

Hygiene factors and motivators

Frederick Herzberg (1959) developed a theory of motivation that identified the following factors as true motivators contributing to high morale and job satisfaction:

- achievement

Motivation and satisfaction in social care

Look at the three figures below from a National Survey of Care Workers completed for Skills For Care (TNS UK Ltd, 2007).

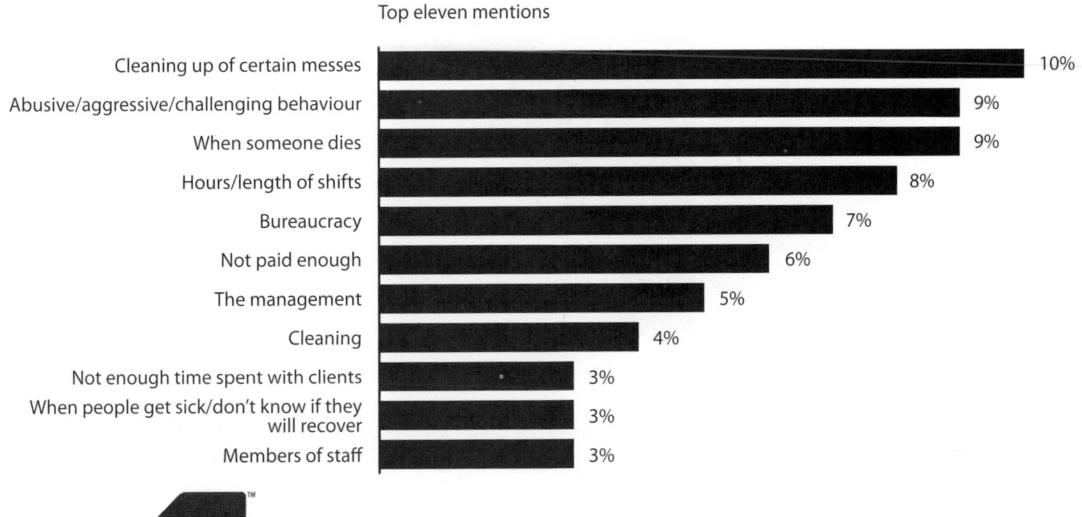

Figure 5.13 The worst things about work

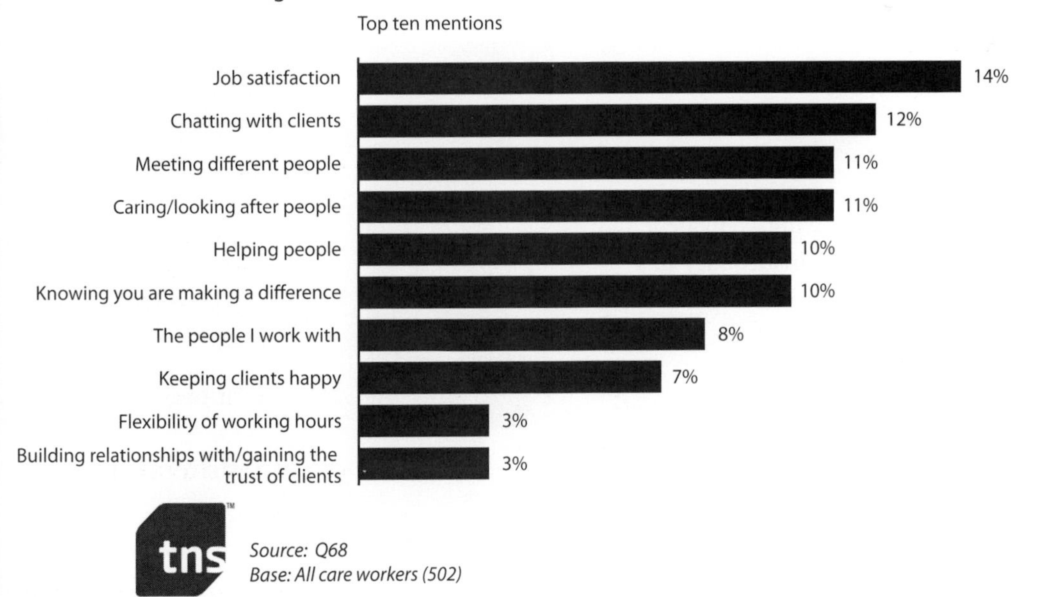

Figure 5.14 Favourite things about work

CASE STUDY continued

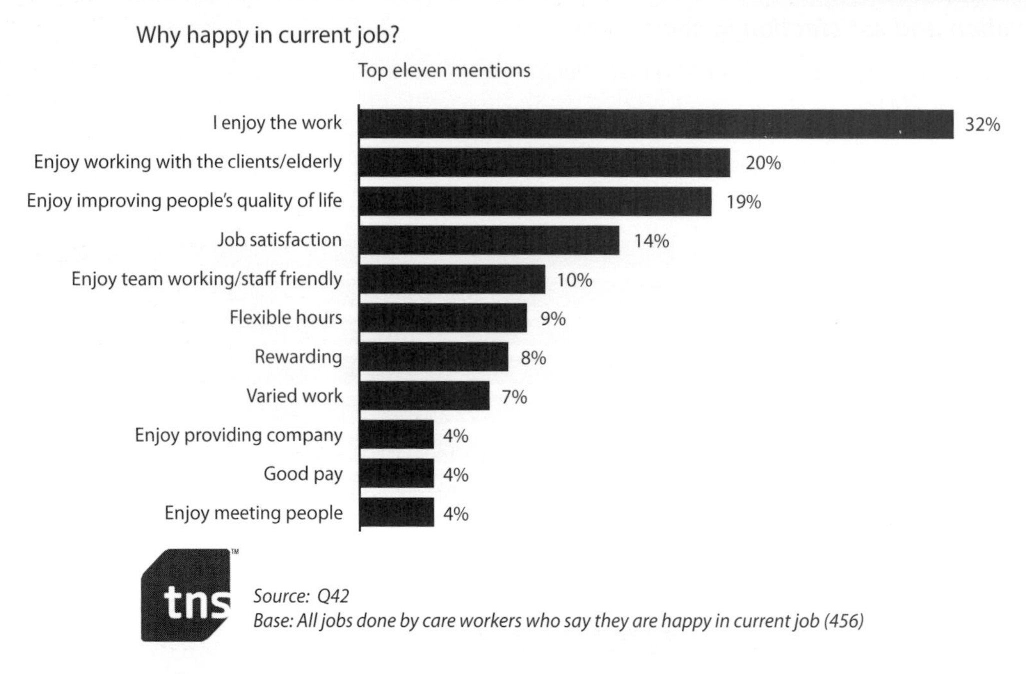

Figure 5.15

Reasons to be happy in current job

- *Do the responses from the survey support McClelland, or Herzberg's theories of motivation?*
- *Which theory, if any, do they support most strongly?*
- *How could the theories and results from the survey inform your practice in motivating staff?*

- recognition
- responsibility
- promotion prospects
- nature of the job.

Herzberg also identified the following as 'hygiene' factors – factors whose absence or inadequacy in a job results in poor performance and dissatisfaction:

- higher authority policy
- pay
- type of management
- working conditions
- relations with colleagues
- fringe benefits.

Herzberg believed that hygiene factors were rarely high (or true) motivators, but were things people tend to expect as normal working conditions.

Figure 5.15 Job satisfaction is very important

Delegation

Delegation involves giving responsibility for particular tasks or areas of work to other staff within your organisation (or within the interprofessional team). People often find it difficult to delegate for a number of reasons:

- because they are anxious that the task will not be carried out as well or as quickly as they can complete it themselves
- because they are unsure which tasks should be delegated
- because they think delegating tasks would undermine their authority
- because they are so busy doing the tasks that they do not take time to identify which tasks could be delegated effectively.

Benefits of delegation

There are numerous benefits to be derived from appropriate delegation, so it worth taking time to review your practice in this area. Delegation:

- improves your time management, allowing you to devote your attention to other management tasks
- helps to develop staff, providing them with the opportunity to increase their skills and knowledge in particular areas. Staff might perform certain tasks more effectively and efficiently than you
- helps to keep staff motivated
- means that staff feel more 'ownership'.

When to delegate

Before delegating tasks, you should consider:

- **the complexity of the task**. If you are unsure about delegating tasks, start with smaller, less vital tasks
- **the previous experience and potential of the member of staff** and whether the task is appropriate for them. Select the best, not necessarily the most convenient person for the job
- **the level of risk involved**. You will be less likely to delegate high-risk tasks completely. For example, you would be unlikely to delegate the whole task of preparing for an inspection of service, but might delegate specific areas or tasks
- **the pressure on you to complete the task** – this should not be the primary reason for delegating.

Love (1997) identifies four distinct styles that managers adopt in delegating (see Figure 5.17).

Style	Attitude of staff	Manager approach
Delegate	Staff member is willing/confident and able	The manager has outlined the task and timescale for completion. They are able to allow the staff to get on with it. Still available during process and review on completion (and mid-way).
Coach/consult	Staff member is willing and confident but lacks ability in relation to the task	The manager needs to provide some coaching and to be available for consultation and regular reviews.
Sell/explain	Staff member is able, but unwilling or lacking in confidence in relation to the task	The manager needs to sell the task and explain how to achieve it. Task may need to be broken down and the manager involved in coaching and supporting the staff member during the process.
Tell	Staff member is unwilling and lacking in confidence and they refuse the task	Manager clearly states what needs to be done, how, by when, etc. This should be the last resort.

(Adapted from Love, 1997)

Figure 5.17 Delegation styles

Also bear in mind that at first delegating a task may **increase** your workload, because you need to spend time supporting and guiding staff. You are not delegating responsibility, and so must still monitor the progress of the task.

REFLECT

Keep a detailed record of your activities over a whole day. Consider each activity in turn and whether you could have delegated any activity. Clearly identify why it was not appropriate to delegate the activity. Where you have identified activities that could have been delegated, how will you plan to do so in the future?

Delegation good practice

- Delegate to the most appropriate person/people: consider their skills, the time they have available, their interests and any equal opportunities issues.
- Don't just delegate tasks that you don't like!
- Make sure that your request and the task is clear (what, when, and what level of autonomy they have).
- Consider if it would be helpful to confirm the details in writing.
- Allow the member of staff (and team where appropriate) to express their view.
- Inform other team members when a task has been delegated.
- Be realistic in your demands.
- Remember: you are still accountable for the work and need to agree how progress will be monitored.

NVQ/SVQ

This section will help to provide evidence for the following element D3.3:

Work Products you may generate and either include in your portfolio or show your assessor to demonstrate your skills and knowledge are:

- **records of supervision, team meetings or other meetings where you have delegated tasks and/or reviewed progress.**

Managing poor performance

Some of the challenges already considered in this chapter may cause a drop in the level of performance of individual members of staff. Here we will consider examples of poor performance, and strategies and procedures for dealing with these. However, it is important to begin by reviewing some of the key strategies (discussed in detail in Chapter 4) that help prevent issues of poor performance arising in the first instance. These are:

- careful recruitment, selection and training
- providing accurate job descriptions
- when an employee starts, explaining the standards of work required, the conditions of any probationary period, and the consequences of failure to meet the necessary standards
- ensuring all staff receive regular supervision and appraisals, and have access to development opportunities
- where there are issues of poor performance explaining to the employee the improvement required, the support that will be given and when and how performance will be reviewed.

As stated earlier, it is best if everyone works towards preventing issues of poor performance arising. However, if you do need to deal with such issues, you may find it challenging, for a number of reasons:

- Managing poor performance is time-consuming.
- It takes confidence and requires support, especially if you are new to your role and/or this is your first experience of poor performance.
- You may encounter pressure from individuals or teams to ignore or accept the poor performance.
- It can raise significant emotions in those involved: for example, you may feel, or be made to feel, responsible or guilty.
- It may feel as if you are involved in an 'uncaring' act in a 'caring' profession.

All the same, issues of poor performance must be dealt with to ensure that:

- service users receive the best possible service, and are not put at risk
- other team members do not have to deal with the consequences of poor performance: for example, 'carrying' the individual(s) concerned
- you do not have to deal with a more serious situation at a later stage
- the image of your organisation and of the social care workforce is not tarnished.

When dealing with poor performance, following organisational policies and procedures is crucial.

VIEWPOINT

List your organisation's policies and procedures for dealing with areas of poor performance. Consider whether you are familiar enough with the policies and procedures to know how to use them. List any other individuals within the organisation who could support and advise you in managing poor performance.

Firstly, to minimise the likelihood of poor performance issues arising within your team, it is important to ensure that all members of staff are clear about the standard of work expected of them.

ACTIVITY

Are standards and expectations clear?

Answer this set of questions provided by SCIE (no date).

1. Are you clear about your own/the agency's/the profession's expectations of:
   - acceptable behaviour at work?
   - performance standards?
   - ways of negotiating differences of professional opinion?
2. Are you confident that these expectations are explicit and shared within the team you manage?
3. How do you tell your service users and carers what standards they can expect from you and your team?
4. Are you confident that you will be told by team members about problems and concerns regarding behaviour and standards?
5. Does the whole team have a stated, shared responsibility for ensuring appropriate standards of behaviour and professional practice: that is, do you collectively safeguard the interests of your service users and carers?

Once you have answered these questions, identify any areas which you need to address more fully. Devise an action plan, with timescales, to help you do so.

'Setting up to fail' syndrome

It is also important that you do not fall into the 'setting up to fail' syndrome – a situation where otherwise effective managers can inadvertently cause good employees to fail (see Figure 5.18). The syndrome was identified by Jean-Francois Manzoni and Jean-Louis Barsoux (2002), and is helpfully explained by SCIE on their People Management website.

In brief, the syndrome occurs where managers form (negative) views quickly and change behaviour accordingly (often unconsciously). This leads them to:

- give an employee more structured and mundane work
- monitor closely and intervene quickly, which the employee often perceives as oppressive.

The manager interprets behaviours that confirm their view: for example, working long hours is an indication of being 'slow' and 'incapable', rather than 'dedicated' and 'motivated'. The employee picks up cues from the manager and begins to fulfil the prophesy: the employee withdraws, feels unable to ask for help/support and so is more likely to make mistakes, and also to try to cover up problems.

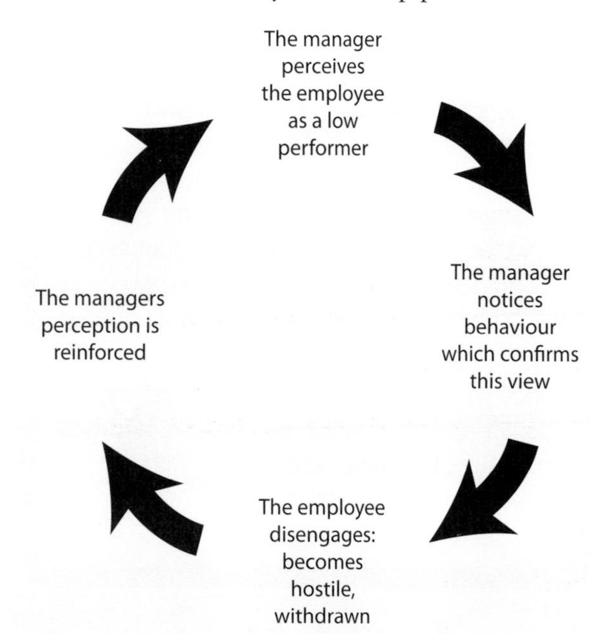

Figure 5.18 'Setting up to fail' syndrome

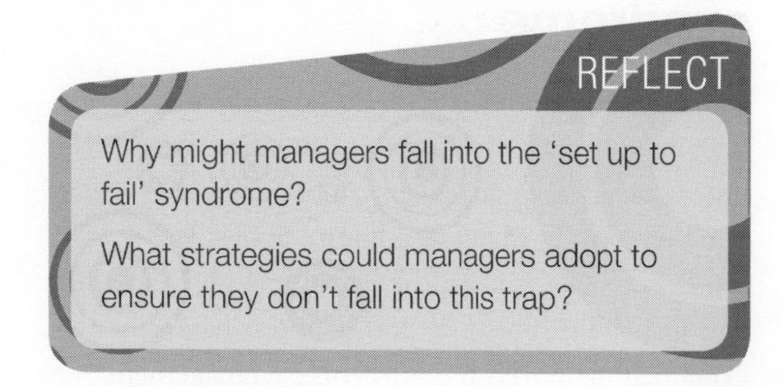

REFLECT

Why might managers fall into the 'set up to fail' syndrome?

What strategies could managers adopt to ensure they don't fall into this trap?

Disciplinary and capability processes

Once you have established that there are issues of poor performance, the questions (Figure 5.19) and checklist from SCIE that follow will help you to consider your next steps. 'Formal action' and 'informal action' are terms used within some organisations' procedures. SCIE stresses that, in this context, 'informal means action that does not invoke a procedure. It does not imply unclear, inconsistent or ambiguous management'. SCIE also reminds us

1. Is your concern about professional behaviour – that is, about how the worker undertakes the tasks and activities particular to their post?

   For example, you may be concerned about:
   - the quality of personal care provided by a care assistant
   - the dependability of the file storage system undertaken by an administrator.
2. Is your concern about 'employee' behaviour – that is, about how the worker fulfils the general requirements of employment?

   For example, you may be concerned about an individual's:
   - lateness
   - discourtesy to colleagues.
3. What level of concern does the behaviour cause you and/or others?

   For example:
   - have you received complaints, formal or informal?
   - is it persistent?
   - have you spoken to the individual but noted no change?
   - is the behaviour specifically covered by agency procedures as requiring formal action?
   - do you think you are dealing with dangerous or unsafe practice?

Figure 5.19 SCIE questions on formal or informal action

that it is important to feel supported and justified both when deciding whether behaviour is unacceptable and when deciding how to manage it, so seek support from appropriate others, such as your line manager.

Here is SCIE's checklist of actions for dealing with poor performance. You should normally ensure that you cover each stage in turn, before resorting to formal action.

Dealing with poor performance

- Investigate fully the reasons for poor performance. ☐
- Discuss with the employee his/her shortcomings and the standard required. ☐
- Provide a reasonable time to improve.
- Advise on how improvement can be made and provision of training and supervision, where appropriate. ☐
- Warn the employee of the consequences of not improving. ☐
- Monitor the employee's performance during and at the end of the 'trial period', while providing the employee with feedback. ☐
- In many cases, improvement will not be instant and, depending on the nature of the job and on the effect of sub-standard performance on the department, the informal process should continue for a reasonable period. ☐
- This process is the concern of the manager and the employee and, although guidance and advice may be sought, representation of the employee by a Union representative or fellow employee is not appropriate at this stage. ☐

Informal action

Where you decide that informal action is the most appropriate approach for dealing with poor performance, you should use supervision to discuss the issues and your concerns, and come to an agreement about a way forward.

SCIE recommends that you:

- try to identify and understand any reasons causing the poor performance
- tell the worker what aspect of their work you are concerned about
- tell the worker why you are concerned
- tell the worker what would be acceptable work
- agree how and by when the worker improves
- agree how you will help the worker to make the changes required.

In dealing with poor performance, you might find it helpful to remember the acronym 'turn on the TAPS' (Strebler, 2004).

Turning on the TAPS

T	**t**imely and early
A	**a**ppropriate management style and response
P	keep it **p**rivate
S	make it **s**pecific to performance, and factual

(Strebler, 2004)

Figure 5.20 Tackling poor performance

Formal action

You should consider formal action where:

- you have tried to manage the poor performance issues informally, and the worker has been given detailed and regular feedback on areas of poor performance, is aware of the areas of concern and understands how to improve their performance, but you have observed no, or insufficient, improvement

- the worker acts in an unprofessional or unsafe way, despite feedback and support
- the worker does not respond, acknowledge or act on feedback regarding their performance, showing lack of insight and/or resistance to change
- the behaviour that concerns you is covered by an organisational policy and procedure.

CASE STUDY

Action planning

Consider these scenarios.

Scenario one *You have heard some of Paula's colleagues refer to her as 'grumpy' on morning shifts on several occasions. Today a resident mentioned to you that Paula had 'really upset' another resident, by speaking to her harshly over breakfast.*

Scenario two *Rosemarie is excellent at interacting with service users and their family members, but is not very organised in her paperwork. She has recently 'mislaid' two important documents, one relating to a service user and the other a receipt for a piece of equipment.*

Scenario three *You have noticed that Brendan has begun to look more untidy, and also tired at work. Recently he has worn clothes that are torn and stained. Today he fell asleep while sitting and talking with a service user.*

For each of the scenarios, answer these questions.

1. *What are possible reasons for the behaviour?*
2. *Are you concerned about the behaviour and, if so, why?*
3. *What changes would you expect to see? And by when?*
4. *How could you support the worker to make the changes required?*

Your organisation's policies and procedures will outline behaviour that would warrant formal action immediately. This will be behaviour of a serious and/or dangerous nature, such as assault or theft, normally termed 'gross misconduct'.

Capability or disciplinary?

Ensure that you are clear whether you are dealing with capability or disciplinary issues. The Advisory, Conciliation and Arbitration Service (ACAS) (2006) defines these areas and provides useful flow charts to show key stages in the processes. You should always seek support from your line manager and Human Resources before and during the implementation of these processes.

Capability issues relate to an employee's ability or qualification to do their job. SCIE gives examples of capability issues, including:

- poor standards of work: for example, frequent mistakes, not following a job through, unable to cope with instructions given
- inability to cope with a reasonable volume of work to a satisfactory standard
- poor attitude to work: for example, poor interpersonal skills, lack of commitment and drive.

Disciplinary issues relate to an employee's misconduct or unsatisfactory performance. SCIE says that examples of misconduct might be:

- refusal or deliberate failure to comply with legitimate management instructions, agency policies or procedures
- unauthorised absence from work
- downloading unauthorised material from websites onto the organisation's computer(s)
- abusive language or behaviour.

Flow chart of the Capability Procedure

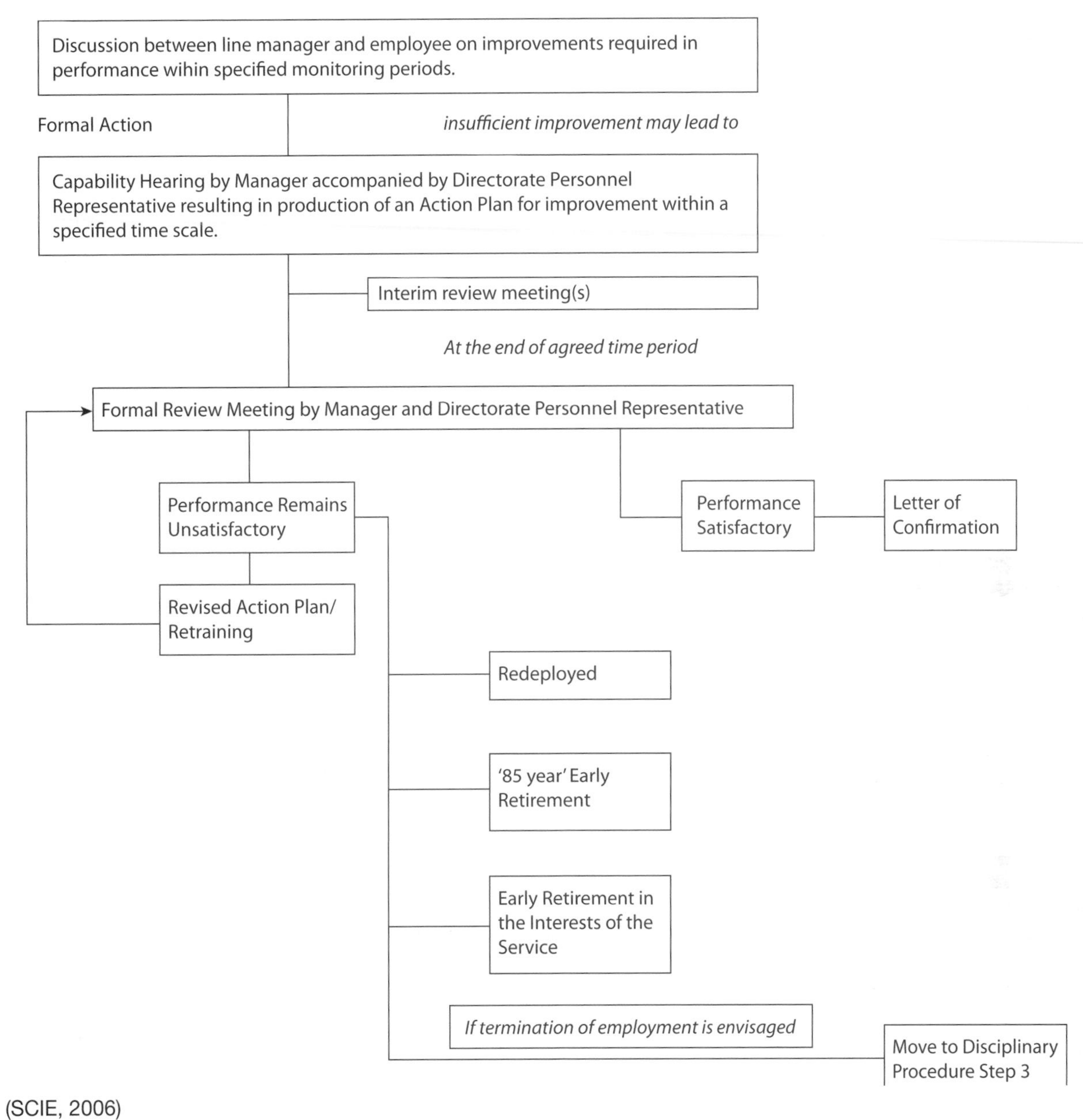

(SCIE, 2006)

Figure 5.21 Capability procedure

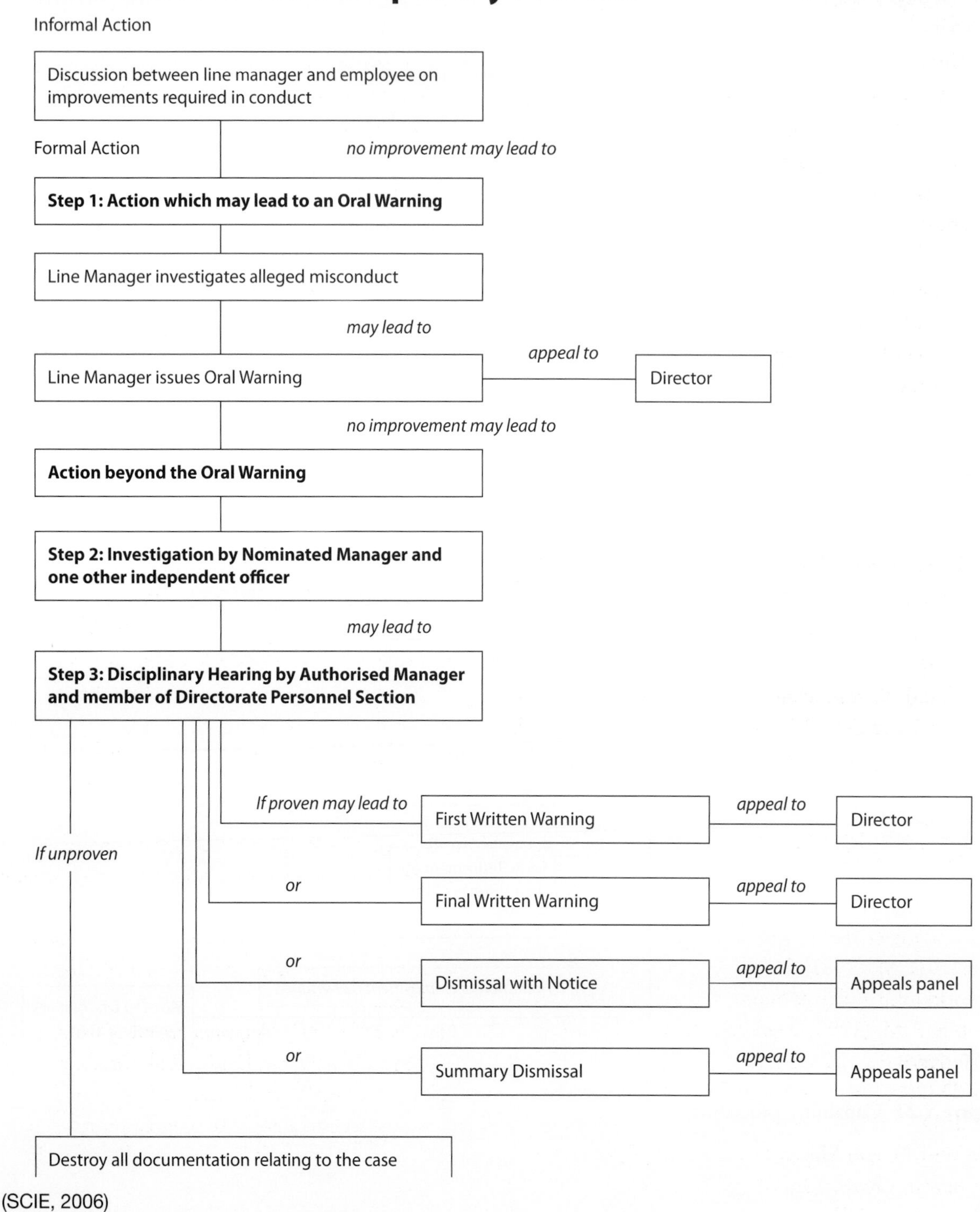

(SCIE, 2006)

Figure 5.22 Disciplinary procedure

NVQ/SVQ

This section will help to provide evidence for units A5, E2, E10.

Work Products you may generate and either include in your portfolio or show your assessor to demonstrate your skills and knowledge are:

- records of any meetings where performance issues have been discussed and recorded
- records of any capability or disciplinary processes you have been involved in.

NVQ/SVQ

This chapter will help you to provide evidence for the following Units:

- mandatory units A1, E1
- optional units A5, B4, E10
- and the following elements from Units E2 and D3 – E2.1 and D3.3

References

ACAS (2006) *Advisory handbook : Discipline and grievances at work*, London: ACAS

Belbin, R. M. (1981) *Management Teams*, New York: John Wiley & Sons

Belbin, R.M. (1993) *Team Roles at Work*, Oxford: Butterworth Heinemann

CIPD (2007) *Managing Conflict at Work*, London: CIPD

De Witt, B. and Meyer, R. (2004) *Strategy: Process, Content, Context* (3rd edition), London: International Thompson

DOH (2008d) *Transforming Social Care*, LAC (DH) (2008) 1

Herzberg, F., Mausner, B. and Snyderman, B.B. (1959) *The Motivation to Work*, New York: John Wiley

Krüger, W. (1996) 'Implementation: The Core Task of Change Management', CEMS Business Review, Vol. 1, 1996

Love, C. (1997) *Developing People The Manager's Role – 20 Tried and Tested Activities for Helping Managers Foster a Positive Learning Climate in the Workplace*, Ely: Fenman

Manzoni, J. and Barsoux, J. (2002) *The Set-Up-to-Fail Syndrome: How Good Managers Cause Great People to Fail*, Boston: HBS Press Book

McClelland, D. (1988) *Human Motivation* Cambridge, Cambridge University Press

Roberts, D., Scharf, T., Bernard, M., and Crome, P. (2007) *Identification of deafblind dual sensory impairment in older people*, London: SCIE

SCIE (2008) *People Management* www.scie.org.uk/workforce/peoplemanagement.asp (accessed 03.05.08)

SCIE (no date) *Practice Guide 1: Managing Practice* www.scie.org.uk/publications/practiceguides/bpg1/index.asp (accessed 03.05.08)

Scott, C. and Jaffe, D. (2004) *Managing Change at Work: Leading people through organizational transitions* (3rd edition), Boston: Thompson Learning

Scragg, T. (2001) *Managing at the Front Line: A handbook for managers in social care agencies*, Brighton: Pavilion

Strebler, M. (2004) *Tackling Poor Performance* (Report 406), Brighton: Institute for Employment Studies

Thomas, K.W. and Kilmann, R.H. (1974) *Thomas Kilmann Conflict Mode Instrument*, California: CPP Inc.

TNS UK Ltd (2007), *National Survey of Care Workers: Final Report*, Leeds: Skills for Care

Tuckman, Bruce W. and Jensen, Mary Ann C. (1977) 'Stages of small group development revisited', *Group and Organizational Studies*, 2: 419–427

University of Victoria (2008), *Team Effectiveness Model* web.uvic.ca/hr/hrhandbook/organizdev/teammodel.pdf (accessed 03.05.08)

Useful Reading

ACAS (2003) *Code of Practice 1: Disciplinary and Grievance Procedures*, London: TSO

ACAS (2006) *Advisory handbook: Discipline and grievances at work*, London: ACAS

SCIE (2008) *People Management* http://www.scie.org.uk/workforce/peoplemanagement.asp (accessed 23.06.08)

CHAPTER 6

Delivering quality services

'It was good because staff and residents made me feel really welcome and they checked to make sure I had everything.'

'Yes, they help me in the right way. They never do anything I don't want them to.'

(Scottish Commission for the Regulation of Care, 2004:17,31)

Introduction

This chapter explores key areas related to the delivery of quality services. This includes the current legal and policy context in relation to quality issues and strategies for evaluating and improving services.

All chapters within this book are related to the provision of quality services. This chapter focuses in more depth on some of the key aspects of quality within the health and social care setting.

The chapter covers:

- definitions of quality
- service-user perspectives
- the legal and policy context
- setting and supporting the quality agenda in your setting
- evaluating services
- improving services.

Definitions of quality

ACTIVITY

Defining quality: Part 1

Think about a service you receive regularly: this might be from an optician, a hairdresser, a garage or a restaurant. List all of the things which you expect from this service as an indicator of good quality.

Now try to write a sentence which sums up your definition of quality.

Would your definition also apply to health and social care services? Are there any elements missing? If so, what and why?

Who defines quality in registered services?

A range of stakeholders is involved in defining quality in registered services. Stakeholders are any individuals or

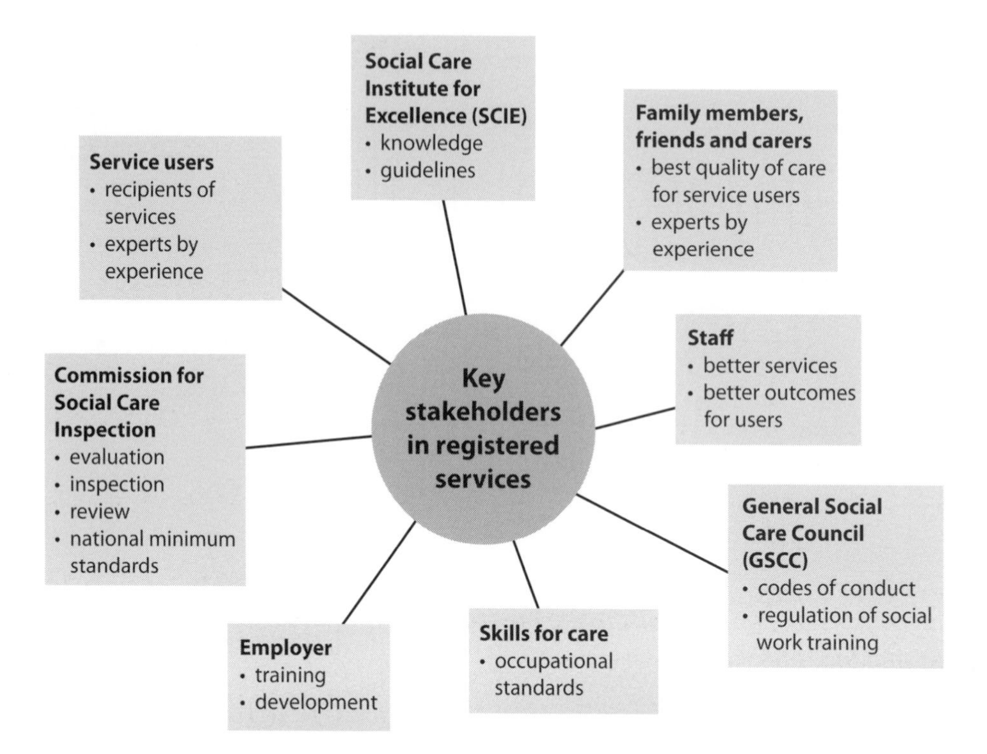

Figure 6.1 Key stakeholders in registered services

organisations that have a direct interest in the service being provided. The individuals and organisations likely to be the key stakeholders in registered services are identified in Figure 6.1.

REFLECT

Think of your own workplace.

- How might the different people above define quality in your service? What elements do you think they would identify?
- Do you think they would define quality in different ways? If so, how? Why do you think they might have different views of quality?

This chapter will focus on quality issues as defined primarily by service users and carers, and also through law and policy. It is vital to focus on the views of service users and carers as they are the recipients of the service, and listening to service users and carers is the central element of a person-centred approach (see Chapter 3). Law and policy are considered as they set the context within which registered services operate, detail the government standards by which quality is measured, and also direct the agenda of many other stakeholders, such as care managers and social workers.

Service-user perspectives

It is important to start from the perspective of those who use services, as recipients or consumers of the service. The current government has stressed that it is crucial that 'social care moves the focus away from who provides the care, and places it firmly on the quality of services experienced by individuals and their carers and families' (DOH, 1998) and that 'all standards will reflect what people say they value in services' (DOH, 2000).

Research conducted by Innes et al. (2006) and Cantley and Cook (2006) found that service users defined quality care as:

- user-focused
- promoting independence and autonomy rather than control
- flexibility and responsiveness
- promoting choice and involvement
- supportive of family members
- services which are non-stigmatising
- services that tend to be offered by those working with a collaborative/team philosophy.

(Innes et al., 2006; Cantley and Cook, 2006)

ACTIVITY

Factors within quality

- Compare your definition of quality and list of factors identified within the first activity in this chapter with the findings outlined above.
- Do they include similar elements? You may have identified differences and/or similarities – why do you think this is the case?
- Do any factors surprise you? Why?

Chapter 3 also addresses a range of areas related to person-centred practice, and you may find it useful to review this.

Finally, the following comment from one senior member of staff illustrates how a care centre in the north-east of England endeavours to ensure that, in providing dementia care, new staff develop responses based on understanding the experience of the service user:

> 'I have had to explain to the staff what it must feel like for the other person – to be in the situation they are in, where the staff were [saying] "Well, you have to see it from our point of view" No, actually, we have to see it from their point of view, and I think that has shifted as well.'

(Cantley and Cook, 2006)

Experts by experience

The term and concept of 'experts by experience' is a relatively recent one, but one which has gained support from government departments, local authorities and other organisations such as the Social Care Institute for Excellence and Commission for Social Care Inspection.

The Commission for Social Care Inspection's definition of an 'expert by experience' is someone with experience of using social care services including:

- people who are using services now or have done in the past
- people who need services but haven't been offered them
- people who need services but haven't been offered any that are appropriate (i.e. culturally), or they are not offered services, (i.e. asylum seekers)
- people living with or caring for a person who uses services.

(CSCI, 2007)

The term 'experts by experience' tends to be used when referring to the direct involvement of service users in the development, delivery and evaluation of services. This has included involvement in areas such as the inspection of registered services (e.g. with CSCI) and in the training of social work and social care staff (see SCIE and GSCC websites for further information). All of these developments ultimately seek to improve the type and quality of health and social care services provided.

Involving service users in the selection of staff

Some service providers involve service users in the shortlisting and interviewing processes when recruiting new staff. This can offer a different perspective within the interview and also help service users to feel more actively involved in the provision and quality of the services they receive.

Involving service users as representatives in management groups or forums

This ensures that a service user perspective is represented in the decision making processes in such meetings and forums. It may be more comfortable and empowering for service users if there is more than one service user representative sitting in the group or forum.

Involving service users in the evaluation of services

Benefits of involving service users in service evaluations include ensuring the service user's perspective is taken when designing evaluations, for example ensuring measures of quality do not just relate to cost or achievement of minimum standards. In addition, people who are currently using services may feel more able to talk and express their views freely.

When involving service users it is important to consider whether it might be more appropriate to involve service users who are currently receiving services, previous service users or service user representatives from organisations such as Mind. You may find it useful to consult with local and national voluntary organisations regarding the services they may be able to offer. It is vital that service users are adequately prepared for any role they play. For example, service users should be given access to the organisations recruitment and selection training. It is also important to ensure that service users are clear regarding areas such as confidentiality. Voluntary organisations may also be able to assist in preparing service users for their role and supporting them in undertaking these roles.

REFLECT

What involvement do 'experts by experience' have in the development, delivery and evaluation of services within your organisation?

How could you improve their level of involvement?

CASE STUDY

Staying user-focused

Brierdene residential home for older people has recently completed a survey of a number of different aspects of the care offered within the home. The area which came out worst in terms of service user satisfaction was mealtimes. Several service users have described it as 'regimented' or 'like a conveyor belt'.

Identify changes you might introduce to mealtimes to overcome this, using the characteristics of quality care as defined by service users in the research by Innes et al. (2006) and Cantley and Cook (2006). For example, how might you make mealtimes more user-focused?

Legal and policy context

Since coming into power in 1997, the government has stressed the importance of raising standards in health and social care services, and this can be seen in a range of legislation and policy introduced over the past decade. The key pieces of legislation and policy and guidance linked to quality issues and standards in health and social care are identified in Figure 6.2.

Law/Policy	Relevance to quality issues
A Quality Strategy for Social Care (DOH, 2000)	*A Quality Strategy For Social Care* (2000) is a document which the government used as the basis of a consultation with people who use and work in social services. It sets out the reforms needed in working practices, local management and training in order to improve the quality of social services in England for the 21st century.
Care Standards Act 2000 or Regulation of Care (Scotland) Act 2001	The Act establishes a system to regulate care services, including the registration and inspection of care providers.
Health and Social Care Act, 2008	This Act establishes the Care Quality Commission, which will merge the Commission for Social Care Inspection, the Healthcare Commission and the Mental Health Act Commission. The Care Quality Commission is expected to become fully operational in April 2009.
Regulations for care homes and care services	Different sets of regulations have been established which apply to different service areas (e.g. Care Homes for Adults or Domiciliary Care Services). There are also sets of regulations that apply across all service types. Regulations are legal requirements that services must meet if they want to be registered providers. Breaches are punishable, including by being barred from providing care.
National Minimum Standards (National Care Standards – Scotland)	National Minimum Standards constitute the minimum expectations the government sets for care providers in the services they deliver. National Minimum Standards are not legally enforceable but are guidelines for providers, commissioners and users to assist them to judge the quality of a service. CSCI inspectors must also take the standards into account.
National Service Frameworks	National Service Frameworks (NSFs) are long-term strategies for improving specific areas of care. They set national standards and identify key areas for intervention, e.g. Preventing Falls (National Service Framework for Older People, DOH, 2001).
Our Health, Our Care, Our Say (DOH, 2007)	This White Paper identifies the seven outcomes the government seeks for people who use services. The performance assessment framework used by the CSCI is built around these seven outcomes (see below).
Codes of Practice for Employers and Employees (General Social Care Council; Care Council for Wales; Scottish Social Services Council, Northern Ireland Social Care Council)	The Care Councils developed Codes of Practice for Social Care Workers and Employers of Social Care Workers. These established for the first time the standards of practice that users of social care services can expect from social care workers, and standards of practice that social care workers can expect from employers.
Single Equality Scheme	Under current legislation public sector organisations are required to promote equality of opportunity and to produce Race, Disability and Gender equality schemes. Many organisations are now opting to develop 'Single Equality Schemes', covering these areas in a single scheme and document. They often choose to also address other areas such as religion, belief, age and sexual orientation. Adopting the 'Single Equality Scheme' approach has the advantage of enabling organisations to identify and respond to issues of multiple discrimination.

Figure 6.2 Law and policy linked to quality issues

The CSCI Performance Assessment Framework (2007b) is built around the seven outcomes that the White Paper on health and social care, *Our Health, Our Care, Our Say* (DOH, 2007), seeks for people who use services:

1. improved health and emotional well-being
2. improved quality of life
3. making a positive contribution
4. increased choice and control
5. freedom from discrimination or harassment
6. economic well-being
7. maintaining personal dignity and respect.

Two additional domains of leadership and commissioning and use of resources are also included separately. These have been added as the CSCI believes that effective outcomes can only be delivered on the back of excellent leadership and effective commissioning and use of resources.

Identify one change or improvement you have made within your service in the last year in relation to each of the seven outcomes. How would you demonstrate this to inspectors?

Supporting the quality agenda

A *Quality Strategy for Social Care* (DOH, 2000) identified a range of people and organisations key to supporting the government's quality agenda.

- The **CSCI** was established in 2004. Its role is to regulate, inspect and review all adult social care services in the public, private and voluntary sectors.
- From April 2009, a new single body responsible for regulating both adult social care and health, the **Care Quality Commission**, will be operational.
- The **Scottish Commission for the Regulation of Care** was set up in 2002 under the Regulation of Care (Scotland) Act 2001 to regulate all adult, child and independent healthcare services in Scotland.
- The **Social Care Institute for Excellence (SCIE)** was established by the government in 2001. SCIE identifies and disseminates the knowledge base for good practice in all aspects of social care throughout the United Kingdom. SCIE comments that 'only by understanding what works in practice – and what does not – can services be improved, and the status of the workforce be raised' (SCIE website, 2008).
- The **Care Councils** were established in 2001. Their main functions are to: establish codes of practice for social care workers and employers; set up registers of social care workers in their country (England, Scotland, Wales, Northern Ireland); regulate social work education and training.
- **Skills for Care (England)** is an employer-led organisation focusing on the training standards and development needs of social care staff in England. Skills for Care collects and analyses data about the social care workforce; develops national occupational standards and qualifications frameworks; promotes new ways of working and delivering services; works closely with organisations such as CSCI and the GSCC.
- The **Department of Health (and Scottish Executive)** is the government department responsible for health and social care policies and standards.

Setting and supporting the quality agenda in your setting

Given the number of stakeholders involved and their potentially different views and expectations, it is not always straightforward to define, deliver and support a quality agenda in your setting. Some key areas to consider for your setting are discussed below.

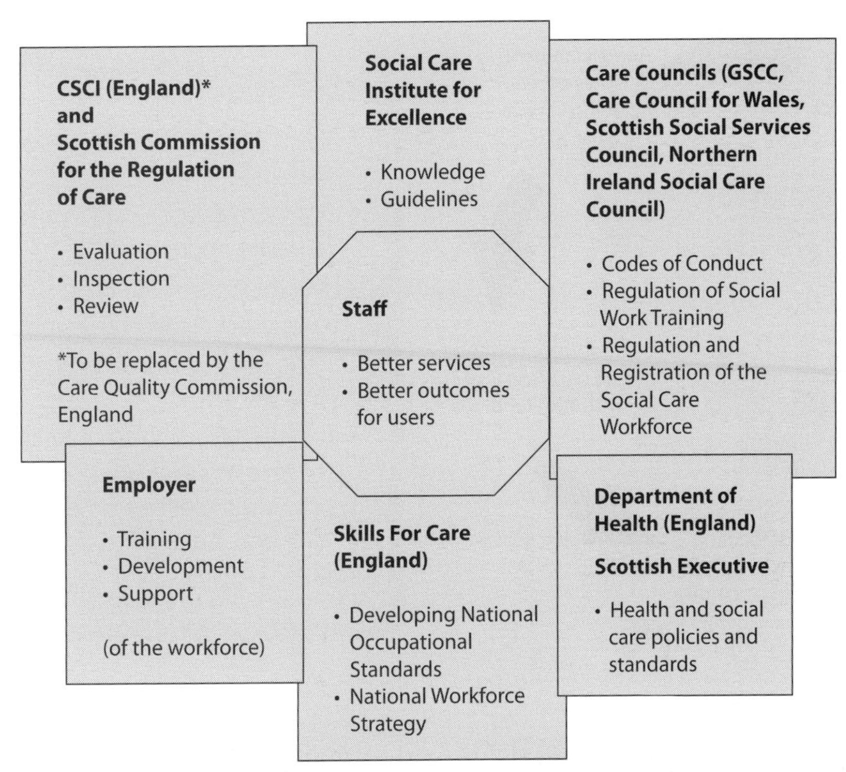

Figure 6.3 Support in the Quality Agenda

Minimum standards or service excellence

It is easy to use minimum standards as the benchmark for a service. However it is important not just to focus on achieving minimum standards but on delivering the best-quality service possible. As a service you may decide to set your own quality standards (either to make standards more specific to your setting or to aim to drive standards above the minimum required). As part of this you may find it useful to go through the following process:

ACTIVITY

Defining quality: Part 2

Think about one of the services mentioned in Part 1 of this activity on page 99.

- What might be the *minimum* standards for these services? For example are the minimum standards for a restaurant that it is clean and the food is edible?
- Do you expect more than the minimum standard from this service? Why?
- Are there situations when minimum standards are good enough and when they are not? When and why?
- Would your answers also apply to the service your organisation/setting provides? Which answers apply, which do not and why?

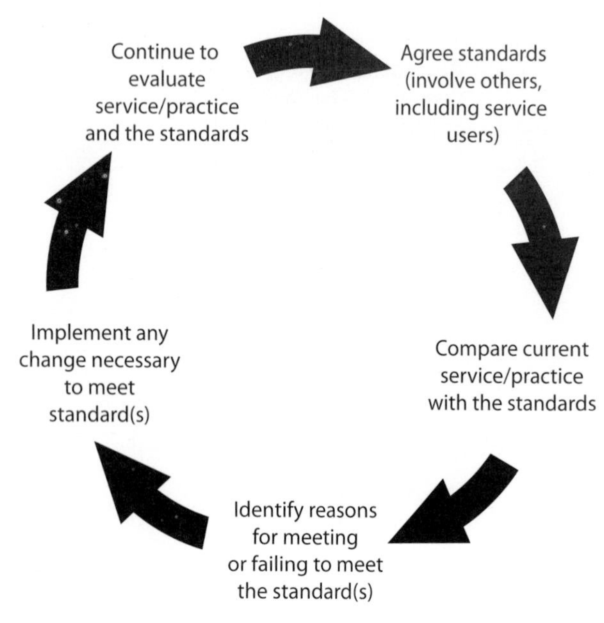

Figure 6.4 Achieving quality standards

REFLECT

- What do you think the strengths of these standards are?
- Are there aspects which are missing or could be improved?
- Would your service meet the standards outlined by Stoke-on-Trent City Council? How would you measure and evidence this?
- Do you have standards written specifically for your organisation or setting? If so, how often are they reviewed and what methods are used to do so? If not, why not?

Agreeing standards

There are a number of factors which need to be considered when setting quality standards.

- They should be developed with the involvement of service users and staff.
- They should be written in plain English and in a form which is meaningful to service users, and should be made available to all service users and staff. Depending on your service user group you may wish to consider the use of visual images, or producing the standards in different formats or languages.
- They should be measurable. Standards have to be set in a way which enables service users, staff and managers to monitor whether the standards are being met.

(Improvement and Development Agency and Audit Commission, 2006)

Examples of standards relating to a number of areas are listed on page 107. These are published by Stoke-on-Trent City Council (2008).

Evaluating services

It is important not to rely only on external inspection by organisations such as the CSCI for evaluation of your services. You should continually evaluate the quality and performance of all aspects of the organisation. By completing the reflection point and activity below, you may identify some of the methods of gathering evidence that will help you to evaluate the quality and performance of your service.

REFLECT

List the different ways that you evaluate the quality and performance of:

- your organisation
- your particular service
- your team
- individuals within the team.

What standards or measures do you use to evaluate quality/performance?

Who defines or writes these standards or measures?

Residential Services for Adults

As a prospective service user you can expect:

- Information about the home is updated regularly and available to make sure that you can make an informed choice about where to stay to live.

As a service user you can expect the following:

- A written contract/agreement with the home which will be written in a user friendly format.
- An individual plan of care which sets out how your health, social and personal care needs will be met in a way which maximised your independence.
- To be supported in maintaining contact with family, friends, representatives and the local community as you wish.
- A room that will suit your needs, be safe, clean, homely and comfortable and, wherever possible, have your own possessions.
- Your needs to be met by adequate numbers of appropriately trained staff.
- Your financial interests to be safeguarded.

In addition to these we comply with the National Minimum Standards for Residential Care for Adults.

Keeping you informed

We are committed to keeping you informed by:

- letting you know how we are performing against our standards
- telling you how we have used your feedback to improve what we do or the way we do things
- advising you about changes we plan to make to our services.

Our equalities statement

We value the variety of our communities and everyone to have their fair access to our services. We are therefore committed to:

- removing discrimination
- promoting equal opportunities
- encouraging good relations with all communities that we serve.

We will take responsibility for these actions by measuring how well we do against the targets we set.

(Stoke-on-Trent City Council, 2008)

Service user forums

Service user forums represent the views of the people directly using the service in question. They may also provide informal support and social networks. Ideally, service user forums should be run by service users themselves.

> 'I was worried about complaining or making suggestions because I didn't want to be a nuisance, but a friend persuaded me that I should. I had a chat with the manager who was really helpful. Now I take part in our residents' and relatives' forum, which is really interesting and I can have my say.'

(CSCI, 2007a:9)

Service user reviews and planning meetings

During reviews and planning meetings, the quality of services being delivered are discussed. It is also a time to review whether the services continue to meet the needs of the service user, identifying what works well and what needs to be developed further.

Complaints and compliments

Complaints can be used constructively to help identify areas of practice, systems or procedures that need to be developed to ensure quality. Similarly, compliments highlight what is working well and reinforce where a good service is being provided. It is important to remember that a low number of complaints does not necessarily mean that a high-quality service is being provided; it may mean that individuals feel unable to make a complaint (for example, because of fear of recrimination). You need to consider how you develop an environment in which individuals feel able to complain.

ACTIVITY

Evaluate your service

This activity is based on the CSCI Annual Quality Assurance Assessment (AQAA) for care homes, which should be undertaken on an annual basis. Complete an evaluation of your services in the areas identified by answering the questions below.

The views of people who use your services

We do the following to ensure that the views of people who use our services are promoted and incorporated into what we do:
We have made the following changes as a result of listening to people who use our services:
We are planning to make these further changes as a result of listening to people who use our services:

Environment

The physical design and layout of the home enables residents to live in a safe, well-maintained and comfortable environment, which encourages independence.

What we do well:
Our evidence to show that we do it well:
What we could do better:
How we have improved in the last 12 months:
Our plans for improvement in the next 12 months:

- Where you have identified areas for improvement, have you included target dates, resources and people who will be responsible for developing the improvements?
- Have you also identified how you will measure when the improvements have been made?
- How would you know if your views about these areas were shared by service users and carers, and also other staff?

Focus groups – staff, service users and/or relatives

Focus groups can look at and give feedback on specific areas of practice, procedures and systems (for example, menu selection). Opinions and feelings can be gathered from individuals in an environment where discussion is actively promoted.

Rabiee (2004) suggests that 6–10 participants within a focus group is manageable, but you may feel comfortable with smaller numbers or smaller numbers may be more appropriate to the needs of the group.

In setting up focus groups you should consider areas such as venue and timing of the group, and diversity of participants. Ensure participants are clear about the purpose of the group. At the beginning of the session it is important to establish ground rules for the group, including ground rules relating to confidentiality and participation within the session. Also ensure that participants are aware that there are likely to be different opinions expressed within the group and that this is fine.

You should prepare key questions before the session and ensure that they are expressed in a way which the participants will understand and are open ended, i.e. will not result in a simple 'yes' or 'no' answer. Also ensure that you check your understanding of the responses – 'Have I understood this correctly?'; 'Does anyone hold a different view?'; 'Is it the same for everyone else?'

At the end of the session explain what will happen next, how participants will be informed of the outcomes and remember to thank participants for their time and effort.

Questionnaires

Questionnaires, which can be in tick-list form, can be a quick and simple method to use when evaluating provision. They can also be anonymous, which may encourage and enable honest opinion from those completing them.

Staff appraisals

Appraisals are a time for staff to evaluate their performance over the last period, and set targets and learning opportunities for the period ahead. This helps staff to identify good practice and areas that need to be developed, which contributes to the quality of the service being delivered. See Chapter 4 for further information on staff appraisals.

Graffiti walls

A wall is cleared for this purpose and normally covered with flip-chart or other paper. There may be different headings or questions on sections of the wall. Participants add their comments with pens or 'post-it' notes, or they may draw or stick pictures to indicate their views and responses. It may be necessary to involve others to help service users to participate and/or to ask service users to explain their contribution (to check that your understanding of their view is correct).

This method can encourage participation as it can be made fun and interactive, and it can also be linked with other events such as social events. The graffiti wall can also be displayed over a period of time allowing service users to contribute when, and if, they wish to do so.

Response cards

Develop sets of response cards such as 'agree' 'disagree' cards, or picture cards for a small group of participants. The participants then hold up the cards in response to the questions posed. You might also ask participants to explain their choice if they wish and if appropriate.

Planning service user involvement

SCIE (2004) provides a useful list of points to consider when planning service user involvement in the improvement of services. Some of these are listed below:

- Be clear about what you want to achieve and any limits there might be (such as time and money) before starting.
- The aims and limits of service user involvement should be clear and understood by everyone involved.
- Before the participation begins make sure that the organisation is prepared to respond to things that service users might suggest.
- Participation should be about service users, not about what managers or other people want.
- Be aware of the relationships and power dynamics between the managers, staff and service users.
- Make sure that information is shared by all.
- People who use services have a very valuable contribution to make. Make sure they are comfortable with the way you are using the information they tell you.
- Be flexible about different ways of working.
- Make sure there is enough time and money to support participation.
- Ask participants how they would like to be given feedback.
- Have a system to see how successful the participation has been and what changes it has made to services. Evaluating the impact will help you to justify continuing to support service user involvement and to improve ways of doing this.
- Talk to service users about how to make sure that everyone who should be represented in the participation is included.
- Think about different ways to involve people who might not usually feel involved in services.
- Make sure all staff understand the value of having service user involvement and have the support to make it successful.

REFLECT

Can you identify any other methods you have used or could use? Identify the strengths and weaknesses of any methods which you have identified.

Improving services

Just as with the evaluation of services, so creativity and innovation in services should be an ongoing process.

CASE STUDY

Reviewing service user involvement

Angela runs Brookfield, a residential care unit for people recovering from mental illness. The primary focus of the unit is to provide a 'half-way' support between hospital and fully independent living in the community. Angela has been asked by her manager to review the level of service user involvement in several areas and to make recommendations about which areas should be prioritised and how involvement might be improved. The areas identified are:

1. *Management and staff meetings and forums*
2. *Recruitment and selection of staff*
3. *Training of staff*
4. *Evaluation of the service*

- *Which areas do you think Angela should prioritise and why?*
- *Are there any areas where it would not be appropriate to involve service users and, if so, why?*
- *Should Angela recommend the involvement of individuals currently using services, or individuals who have previously used this or similar services? Why?*
- *Identify key tasks for Angela in beginning to increase the involvement of service users.*

CASE STUDY

Consulting on improvements

Ashford House is a residential home for people with learning difficulties. It also has four semi-independent flats on its site. A recent CSCI inspection has identified that the range and type of activities offered need to be improved to better meet service users' and carers' needs and wishes.

- *What methods would you use to consult with service users and carers?*
- *Would you use the same or different methods to consult with service users and carers? Why?*
- *Why have you chosen these methods? What are their limitations?*
- *How could you ensure that it was not just the views of the more vocal service users and carers that were heard?*
- *If a wide range of new activities had been identified through the evaluation, and you were only able to offer a limited number of these, how would you select which activities were offered?*

Defining creativity and innovation

The terms 'creativity' and 'innovation' are often used interchangeably, but there is a difference between the two. In straightforward terms, **creativity** is thinking up new ideas from scratch, and **innovation** is implementing the new ideas, or improving on existing systems.

REFLECT

- How can you encourage creativity in your setting?
- How might you help staff to feel more able to suggest new ideas?
- How can you develop your own creativity?

Innovation in your area

Examples of innovation might include:

- introducing a new respite care service
- developing links with the community in which your service is based by linking with a local school and developing a range of projects/activities together
- making mealtimes more interesting by introducing themed nights such as Caribbean or Spanish evenings
- broadening activities; for example, starting a film club
- improving the service you provide to carers and relatives by introducing an evening surgery so you are more accessible to those who have daytime commitments.

ACTIVITY

Innovation and responsibility

Think of an aspect of innovation you have led in your area of responsibility and answer these questions.

- Which aspects of your service needed improvement and innovation?
- What criteria or methods did you use to decide which aspects of the service needed improvement?
- Which aspects of the service did you decide should stay the same and why?
- Who else did you involve in this process and why?
- How did you ensure that those involved were committed to the innovation or changes?
- How did you implement the innovation?
- How did you monitor the process of implementing change and the progress made?
- Was the innovation a success? How did you measure its success?

NVQ/SVQ

This chapter will help you generate evidence for the following units:

- mandatory units A1, B1, C1, E1
- optional units A2, A4, A5, B2, E3

Work Products you may generate and either include in your portfolio or show your assessor to demonstrate your skills and knowledge are:

- evaluations of services or specific areas of practice you have been involved in
- examples of innovations and developments made following evaluations
- minutes of meeting and forums
- questionnaires you have devised to seek feedback from service users, carers or staff
- evidence of research you have undertaken into new services or ways of working, and how you have decided whether and how you will apply this to your setting
- records and reports relating to complaints
- inspection reports that refer specifically to you
- feedback from external organisations or professionals.

CASE STUDY

Night-time care

The following activity highlights a real example of innovation in care services.

A study into the night-time care experiences of residents, relatives and staff in three care homes in Scotland (Kerr et al., 2008), highlighted areas in this aspect of provision which could be improved.

Here are some of the problems that were highlighted.

- *The levels of noise and light during the night were too high to support good sleep for residents. The noise was caused by: staff talking; staff activities; buzzers/resident alarm systems; residents; the fabric of the building.*
- *Routine, indiscriminate overchecking led to unnecessary disturbance of residents. One resident said: 'I am fast asleep and then they open the door and put on the light and I jump awake, my heart jumps and then I cannot get back to sleep.'*
- *Night staff were less well trained and less managed and supervised. They sometimes felt isolated and experienced high levels of anxiety – such as about what would happen if there were a fire or if someone needed medical attention.*

- *What improvements might you introduce to overcome these difficulties?*
- *Why?*
- *Who else would you need to involve?*

The innovations actually introduced by the care homes within the study are outlined in Appendix 1.

In conclusion, high-quality care settings have a significant impact on services users' lives. Providing a high-quality service is about recognising and promoting the areas where you do well and identifying and acting on the areas that need to be developed. Service user feedback is an important and valuable way to measure quality. All managers and settings should be striving for comments from service users like these:

> 'They are very fair, they offer quite a lot of activities and if you don't want to go you don't have to... Most of us really feel that we belong. They do keep us pretty well informed about what is going on. They take into consideration our needs, I find it very fair. They are trying to find out what we need and we are trying to give them some ideas of what we like doing..'.
>
> '... there is no better than the best.'
>
> '... An excellent place, they've just about thought of everything ...'
>
> (Cantley and Cook, 2006)

Figure 6.5 Talking to service users and getting their feedback is an excellent way to measure quality

References

Cantley, C. and Cook, M. (2006) *A report on the evaluation of Moor Allerton Care Centre, Northumberland*, Manchester: Dementia North Centre

CSCI (2007) *People who use services and experts by experience*, www.csci.org.uk

CSCI (2007a) *Care Homes For Older People: National Minimum Standards*, London: CSCI

CSCI (2007b) *Social Services Performance Assessment Framework Indicators Adults 2006–07*, London: CSCI

DOH (1998) *Modernising Social Services*, London: TSO

DOH (2000) *A Quality Strategy for Social Care*, London: TSO

DOH (2003) *Care Homes for Older People: National Minimum Standards*, London: TSO

DOH (2007b) *Our Health, Our Care, Our Say*, London: TSO

Improvement and Development Agency and Audit Commission (2006) *Managing Quality*, www.makingendsmeet.idea.gov.uk

Innes, A., Macpherson, S. and Mccabe, L. (2006) *Promoting person-centred care at the front line*, York: Joseph Rowntree Foundation/SCIE

Kerr, D., Wilkinson, H. and Cunningham, C. (2008) 'Supporting older people in care homes at night', York: Joseph Rowntree Foundation

Rabiee, F (2004) Focus-group interviews and data analysis: Proceedings of the Nutritional Society. 63: 655–660

SCIE (2004) 'SCIE position paper 3: Has service user participation made a difference to social care services?' London: SCIE

Scottish Commission for the Regulation of Care (2004) *A review of the quality of care homes in Scotland 2004* www.carecommission.com/images/stories/documents/publications/reviewsofqualitycare/197.pdf (accessed 14.08.08)

Stoke-on-Trent City Council (2008) Adult Social Care Quality Standards. www.stoke.gov.uk/ccm/content/ss/adult-social-care-quality-service-standards.en (accessed 14.08.08)

CHAPTER 7

Safeguarding vulnerable adults

'Much of it is so hidden – that's the whole point really. People in care homes are victims of the system and it's also the system that has the power to address the problem and resolve it. No one else can really do this.'

(Penhale et al., 2007:153)

Introduction

The Commission for Social Care Inspection highlighted in 2008 that the abuse of vulnerable adults is 'still not universally recognised, understood or accepted as a problem'. This chapter focuses on safeguarding vulnerable adults and covers these key topics to help you to fulfill your responsibilities in this critical area:

- definitions of the terms 'vulnerable adults' and 'abuse'
- categories of abuse
- signs and indicators of abuse
- legislation and policies that underpin safeguarding vulnerable adults
- preventative action
- institutional abuse, including lessons from inquiries and serious case reviews
- safeguarding procedures and processes
- responding to disclosures or alerts of abuse
- whistleblowing.

Definitions

There is no common understanding, or legal definition, of what 'safeguarding adults' means; nor is there any universally accepted definition for the key terms that apply to situations of safeguarding adults: 'safeguarding', 'abuse', 'harm' and 'vulnerable'.

The Department of Health defines a vulnerable adult as a person who:

> 'is or may be in need of community care services by reason of mental or other disability, age or illness; and who is or may be unable to take care of him or herself, or unable to protect him or herself against significant harm or exploitation'.

(DOH and Home Office, 2000:8–9)

One criticism of this definition may be that it seems to identify particular groups of people as vulnerable, such as older people or people with learning disabilities.

Does this definition seem to support a medical or a social model of vulnerable adults?

From the social model perspective, a definition would emphasise more clearly that it is not the adult who is vulnerable, but the situation in which they find themselves. For an explanation of the social and medical models, see Chapter 3 (page 26).

The Department of Health defines abuse as:

> 'a violation of an individual's human and civil rights by any other person or persons. Abuse may consist of a single act or repeated acts. It may be physical, verbal or psychological, it may be an act of neglect or an omission to act, or it may occur when a vulnerable person is persuaded to enter into a financial or sexual transaction to which he or she has not consented, or cannot consent. Abuse can occur in any relationship and may result in significant harm to, or exploitation of, the person subjected to it.'

(DOH and Home Office, 2000:9)

The definition of abuse used by Action on Elder Abuse emphasises the harm and distress caused by abuse:

> 'A single or repeated act or lack of appropriate action, occurring within any relationship where there is an expectation of trust, which causes harm or distress to an older person.'

(Action on Elder Abuse, 2006:1)

Both the Department of Health and Action on Elder Abuse highlight that abuse may be an action or an omission to act. An omission might be ignoring medical or physical care needs or failure to provide access to appropriate health or other services. This links with the 'duty of care' placed on individuals and organisations in the health and social care sector.

Independence, choice and risk (DOH, 2007) defines a duty of care as 'an obligation placed on an individual requiring that they exercise a reasonable standard of care while doing something (or possibly omitting something) that could foreseeably harm others'.

A CSCI (2008) consultation *Raising Voices: views on safeguarding adults* highlights that the range of definitions used by different individuals and organisations can contribute to a lack of clarity of roles and responsibilities in responding to situations where adults need assistance to stay safe. The consultation confirms that people found this situation confusing and unhelpful.

VIEWPOINT

Do the policies and procedures in your setting define terms such as 'safeguarding', 'abuse', 'harm' and 'vulnerable'?

If so, can you identify the source of these definitions? Do they come from government policy documents or are they specific to your organisation or setting?

How, when and where do you ensure that staff in your setting have an opportunity to discuss their understanding of these terms?

Categories of abuse

Abuse can take different forms, and is often divided into a number of categories. It is important to remember that an individual or group may experience one or several forms of abuse, and abuse may fall into more than one category.

These are the categories of abuse that are most commonly listed.

- **Physical abuse** occurs where physical harm has occurred or was intended. This includes hitting, pushing, slapping, kicking, misuse of medication and inappropriate restraint.
- **Sexual abuse** is any sexual act or contact which an individual has not consented to, or where an individual was pressured into consenting. Situations where an individual is unable to consent due to communication difficulties or lack of capacity to give consent are also defined as

sexual abuse. This includes rape and sexual assault or other sexual acts. Forcing an individual to look at pornography or witness sexual acts is also abusive.

- **Psychological or emotional abuse** is abuse that causes harm or distress to the individual's psychological health, development and well-being. Where individuals have suffered another form of abuse such as physical abuse, they are also likely to have emotional abuse as a consequence. Emotional abuse may include shouting, swearing, insulting, threats of harm or abandonment, deprivation of contact, isolation, humiliation, blaming, intimidation, or depriving the individual of privacy or quiet.
- **Financial abuse** relates to the theft or misuse of property and belongings as well as finances. It may include fraud, exploitation, pressure in connection with wills or other documents and processes related to financial transactions, or the misappropriation of benefits.
- **Neglect** links with acts of omission, which are referred to above. Examples might include failure to offer sufficient or appropriate food or drink, failure to provide a safe and appropriate environment (with adequate heating and hygiene levels).
- **Discriminatory abuse** is abuse that has taken place based on an individual's ethnicity, disability, gender, sexual orientation, religion or age. Examples include racist or sexist abuse, other forms of harassment, slurs or similar treatment.
- **Institutional abuse** refers to practices and behaviours that can become common or 'accepted' within an institution such as a residential home, day centre or hospital ward. Institutional abuse often occurs when the smooth running of the institution takes priority over individuals' needs and wishes. Institutional abuse often takes place in a context of rigid routines and the dehumanisation of individuals receiving care. This category of abuse is explored in more depth later in the unit as it is an area which has been highlighted repeatedly within inquiry reports into the abuse of adults who were in vulnerable situations.

CASE STUDY

Is this abuse?

Consider the following situations and answer the questions below.

Renuka *is a vegetarian because of her religious beliefs. During a stay in a residential home for respite care, the staff ignored this and so Renuka eats very little during her stay.*

When they are short staffed, which is often, staff at the ***Willows Care Home*** *put incontinence pads on the residents so that they do not have to take them to the toilet as often.*

The manager at ***Everdene Residential Home*** *has installed cot sides on most of the beds in the home. She says it helps to stop residents wandering and so 'keeps them safer at night'.*

The staff responsible for administering medication give medication due at night-time at teatime instead so that residents can be put to bed at 7 p.m.

Sam*, a senior care worker, tells Jim, one of the residents, that if he doesn't 'behave', he will 'have to go' and the next home will be much worse.*

Jean*, a care assistant, does some shopping for residents, buying newspapers and small items. She keeps the change because it is too much hassle to do all of the paperwork and the residents 'wouldn't know the difference anyway'.*

In each case, ask yourself these questions.

- *Is this abuse?*
- *If not, why not? If it is, which category of abuse would it fall into?*
- *Does it make any difference whether the abuse is intentional?*
- *If you were aware of this situation, would you report it? Why?*

Signs and indicators of abuse

The abuse of adults, like that of children, is often not obvious or visible to others. Staff, family members or friends may be alone with an individual at particular times, or an individual may have difficulties in communicating what is happening to them, or may not have the capacity to understand that they are being abused. It is vital that all staff remain alert to possible signs or indicators of abuse.

Legislation and policies that underpin safeguarding vulnerable adults

A range of legislation and policies underpin the work to safeguard vulnerable adults. However, these laws and policies have developed sporadically over a number of years and so do not always appear streamlined or coordinated.

Key legislation and policies are outlined in the table opposite.

ACTIVITY

Signs and indicators

Complete the table below, identifying signs and indicators that would alert you to possible abuse of an individual within your setting

Category of abuse	Signs and indicators
Physical	
Psychological/emotional	
Sexual	
Financial	
Discriminatory	
Institutional	

Are there some signs and indicators that are common to more than one type of abuse?

Even where one or more indicators are present, it does not necessarily mean that abuse is occurring: there may be other causes. However, it is important to remain alert and open to the possibility of abuse occurring, and to take time to listen to the adults in your care.

The Care Standards Act 2000 and Regulations made under the Act	The Act introduced requirements on care providers to: • ensure they have proper procedures in place to protect people in their care from the risk of actual harm or abuse • to check the Protection of Vulnerable Adults (POVA) list as part of recruitment procedures (from October 2009, POVA checks will be replaced by checking membership of the Independent Safeguarding Authority scheme – see the section below). Under the regulations, the registered person shall ensure that no service user is subject to physical restraint unless restraint of the kind employed is the only practicable means of securing the welfare of that or any other service user and there are exceptional circumstances. (See also Mental Capacity Act 2005: Deprivation of Liberty Safeguards).
Safeguarding Vulnerable Groups Act 2006	The Act is due to be implemented in 2009 and defines who is vulnerable in terms of settings in which a person resides; the types of services a person may receive and whether advocates/representatives have been appointed under various legislative measures. This Act replaces the provisions of the POVA list, the vetting and barring scheme, which was implemented under the Care Standards Act 2000. Under the Act, the Independent Safeguarding Authority (ISA) will assess each individual who wants to work as an employee or volunteer with vulnerable people. Potential employees and volunteers will need to apply to register with the ISA. The ISA will operate across England, Wales and Northern Ireland, with Scotland having a similar system.
Adult Support and Protection (Scotland) Act 2007	This Act defines adults at risk of abuse, places a duty on councils to investigate suspected abuse of a vulnerable adult, and provides powers to intervene in the affairs of adults, provided this is the least restrictive action and is of benefit to the person. Providers are expected to cooperate with the new adult protection systems and be aware of the local multi-agency policies.
Mental Capacity Act 2005	This introduced two new legal offences of mistreatment and willful neglect in respect of people who are thought to 'lack mental capacity'. **Deprivation of Liberty Safeguards** These safeguards come into effect in April 2009, and only apply to people who are assessed as lacking capacity and who live in a residential or nursing home (and are not subject to the Mental Health Act 2007), and are being deprived of or severely restricted in their freedoms. The authorisation to deprive someone of their liberty should be requested by the care home (or hospital) in which the person is resident. If authorised, there is a maximum of 12 months duration of the authorisation (but an expectation that it will be much less). The Act defines restraint as 'the use or threat of force to help do an act which the person resists, or the restriction of the person's liberty of movement, whether or not they resist. Restraint may only be used where it is necessary to protect the person from harm and is proportionate to the risk of harm.'

Continued overleaf

The Sexual Offences Act 2003	The Act relates to people 'with a mental disorder impeding choice', as well as children. Particular sections specify offences involving care workers, abusive types of relationships and abuse of trust.
Human Rights Act 1998	Article 3 prohibits 'torture and inhuman or degrading treatment'; Article 5 acknowledges that 'everyone has the right to liberty and that it should only be restricted if there is specific legal justification'; and Article 14 outlaws 'discrimination of all types'.
Health and Social Care Act 2008	This Act strengthens the protection of people using residential care by ensuring that any independent sector care home that provides accommodation together with nursing or personal care on behalf of a local authority is subject to the Human Rights Act 1998.
Public Interest Disclosure Act 1998	This Act was implemented in 1999, and is to protect workers in the public, private and voluntary sectors from victimisation in employment following a disclosure in a number of circumstances including: a danger to the health and safety of any individual, a criminal offence and the breach of a legal obligation. (See the section on whistleblowing later in this chapter.)
No Secrets (DOH, 2000) *In Safe Hands* (National Assembly for Wales, 2000)	This is Department of Health guidance on developing multi-agency policies and procedures to protect vulnerable adults from abuse. Social services departments were given the lead role in ensuring that local multi-agency policies and codes of practice were developed and implemented by 31 October 2001.
Safeguarding Adults: A National Framework of Standards (ADSS, 2005)	The Association of Directors of Social Services, with a number of partners, published a set of standards for social services departments to use. The aim of the framework is to try to support the work already under way and also to try to reduce the variation in practices and procedures across the country. The standards are not obligatory, but have been adopted by many councils and their partners.

Figure 7.1 Key legislation and policies that underpin safeguarding vulnerable adults

Preventative action

Ideally we would wish to prevent abuse from occurring in the first instance, and managers play a key role in this respect. The CSCI identifies a number of factors that minimise the likelihood of abuse taking place.

- **A culture of openness and dignity** – This enables individuals living in care homes, their families and staff to raise any concerns about abuse.
- **Visible and well understood complaints procedures** – Service users and staff need to be aware of their rights and understand how to make a complaint.

  'Understanding your rights so that you do not feel intimidated or coerced into doing something you need not do, or treatment you may refuse.'

  A person using care services talking about what would help keep them safe

(CSCI, 2007c)

- **Clarity regarding the roles and responsibilities of managers and staff once concerns or complaints have been raised –**

  'Good procedures – but procedures do not keep people safe – the way they are understood, implemented and checked could.'

  A person using care services talking about what would help keep them safe

(CSCI, 2007c)

- **Training in adult protection for all staff** – This forms part of the Care Homes Regulations and

ACTIVITY

Evaluation

Leicester City Council and partners identify a range of measures that service providers should have in place to prevent the abuse of vulnerable adults. Evaluate your setting against these areas. Where you assess the setting as 'Do well', give specific examples to support your view. Where you assess the setting as satisfactory or poor, identify key action points and timescales to improve performance.

Area of practice	Do well	Satisfactory but needs improvement	Poor
Rigorous recruitment practices that include CRB and reference checking			
Thorough induction process including adult protection issues			
Staff awareness of the policy and procedures for protecting vulnerable adults from abuse			
Staff encouraged to be vigilant and to report all concerns			
Culture of openness and transparency			
Open dialogue between staff and managers and service users			
Clear service standards			
Staff aware of standards and expectations			
Staff training			
Person-centred approach to care and support			
Clear complaints procedure accessible to service users, relatives, staff and public			
Staff aware of anti-discriminatory practice			
Regular, recorded, formal staff supervision			
Staff aware of their employment rights support systems: e.g. conditions of employment (including annual, sick and special/carers leave); professional organisations; European Work Time directive			
Risk assessments completed for: assessment of vulnerability and risk of abuse; moving and handling; threats or assault by service users or public; working in a stressful environment			
Comprehensive policies, procedures and staff training on: Challenging behaviour; Personal and intimate care; Physical intervention – in line with new codes of practice; Sexuality and relationships; Medication; Financial accountability; Risk assessment and management; Cultural awareness; Disability awareness; Manual handling; Communication			
Effective monitoring and auditing to ensure quality			
Integration of adult protection into all aspects of care and support			

(Leicester City Council, 2004)

National Minimum Standards. Training will help staff to be aware of indicators of abuse and understand the settings policies and procedures and their role in these. Indeed, findings from inquiries and serious case reviews, such as the investigation into the provision of services for people with learning disabilities at Cornwall Partnership NHS Trust (Healthcare Commission and CSCI, 2006a), highlight the part that lack of training has played in enabling abuse to take place and remain unchallenged.

REFLECT

- Are there any other factors that can help to minimise the likelihood of abuse taking place?
- What would you add to the list and why?
- Compare your list with the items identified by Leicester City Council on page 121 and complete the evaluation exercise.

Institutional abuse

In the 1960s, Goffman (1963) highlighted how processes and methods of working by staff in institutions could adversely affect the people they were supposed to serve. Wardhaugh and Wilding (1998) use the term 'corruption of care' to describe when systems and patterns of organisations can lead to abuse. Institutional abuse often takes place in a context of rigid routines and the dehumanisation of individuals receiving care: for example, people may be referred to by their bed- or room-number rather than by name. Factors and practices which might contribute or lead to institutional abuse are:

- rigid routines
- lack of service-user choice and consultation
- lack of personal belongings
- staff being 'task-focused'
- poor staff morale and high stress levels
- poor awareness, or lack of, policies and procedures
- lack of training.

Lee-Treweek (1998) gives an example of practices where the routines and practices of staff have become more focused on their needs and the needs of the organisation, than on those of the service users.

> 'Morning work was virtually all bedroom work... It was customary to present the patients to the new shift intact, clean and quiet in their rooms for 8 am. Presenting well-ordered bodies seemed to symbolise the job properly done.'

(Geraldine Lee-Treweek in Allott and Robb, 2004:231)

Here are four examples of serious incidents involving institutional abuse in learning disabilities services.

The Independent Longcare Inquiry
(Burgner, 1998)

In 1994, a leaked Buckinghamshire County Council report revealed that, for more than 10 years, former social worker Gordon Rowe had been beating, raping and neglecting the adults with learning difficulties who lived in the residential homes run by his company, Longcare. Other members of staff were also involved in this abuse.

Complaints from staff instigated an investigation by Buckinghamshire's inspection unit but it took nearly 10 years for staff to make those complaints. The inquiry found that they failed to do so earlier primarily because:

- there was an atmosphere of threats and intimidation, and back-biting and bitchiness among staff was rife
- Rowe often told inexperienced care assistants that the violent 'techniques' they saw him use were accepted practice.

A member of staff commented that 'the atmosphere was like walking into a factory. You couldn't wait to get out. I just feel sorry that I didn't do anything, that we weren't strong enough. But I think all of us were worried about what would happen to us.'

(Pring, 2005)

Investigations into Scottish Borders Council and NHS Borders Services for People with Learning Disabilities (Scottish Executive, 2004)

In 2002, a woman was admitted to Borders General Hospital after she had gone to the house of a friend who found her to be badly injured and called an ambulance. When taken to hospital, she was found to have multiple injuries from physical and sexual assault. A police investigation revealed a catalogue of abuse and assaults over the previous weeks and possibly much longer. Three men were convicted of the assaults later in 2002.

The woman was considered to have a learning disability. A series of events had led to her being cared for by one of the convicted offenders. Despite serious concerns about this male's behaviour toward this woman over several years, no action had been taken.

Other individuals were found to be receiving care under similar circumstances. The individuals had varying degrees of learning disabilities, physical disabilities and mental health needs, which were largely neglected, to the point of becoming potentially life-threatening for some. Again, health and social work records contained numerous statements of concern about their care, including allegations of serious abuse and exploitation that were not acted upon. The individuals were found to have been neglected, living in unsuitable and unsanitary conditions and were financially and sexually exploited. The investigations highlighted a catalogue of failings including:

- the failure to investigate very serious allegations of abuse appropriately
- an acceptance of the poor conditions in which the people involved lived and the chaos of their lives
- lack of, or very poor quality, needs assessments and care plans
- lack of information-sharing and co-ordination within and between key agencies
- disagreements between agencies at frontline and middle management level, with no mechanism for resolving these
- unsustained contact with the individuals by the specialist Learning Disability service
- failure by some members of the Primary Care Team (GPs and District Nurses) to act on information about poor home conditions and to make these concerns known to the social work service
- lack of risk assessment and failure to consider allegations of sexual abuse
- very poor standards of case recording
- lack of understanding of the legislative framework for intervention and its capacity to provide protection
- failure to understand and balance the issues of self-determination and protection
- failure to protect the finances of vulnerable individuals
- inability and/or unwillingness to confront aggression and staff's consequent collusion with aggressors to the detriment of victims
- lack of understanding of the complexities of child/adult protection and of the need to explore all allegations of abuse and the possible reasons for retraction of these
- failure to communicate with service users or to engage them effectively in assessing their needs
- lack of compliance with procedures
- infrequent, unstructured and poorly recorded supervision of frontline staff by managers
- serious deficiencies in training and development
- lack of clarity of roles and reporting responsibilities
- uninformed and inaccurate assumptions of individual staff expertise in particular areas and consequent dangerous reliance on this
- lack of senior management and leadership
- ineffective management of poor practice
- breaches of the Scottish Social Services Council Code of Practice for employers.

Joint investigation into the provision of services for people with learning disabilities at Cornwall Partnership NHS Trust

(Healthcare Commission and CSCI, 2006)

In 2005, the Healthcare Commission and the Commission for Social Care Inspection (CSCI) investigated services for people with learning disabilities provided by Cornwall Partnership NHS Trust. The investigation was generated by East Cornwall Mencap Society in October 2004, which raised concerns about the care and treatment of people living in the Trust's assessment and treatment centres and supported living services.

The investigation found that a number of individuals had suffered abuse including physical, emotional and environmental abuse. It found that 'institutional abuse was widespread, preventing people from exercising their rights to independence, choice and inclusion'.

Examples of the practices were:

- unacceptable restrictions on service users: for example, some internal and external doors were kept locked by staff to restrict the movement of people from the services
- poor management of the finances of people in supported living services, such as the apparent pooling of their money to a shared household account and the use of people's money to purchase communal goods and pay for improvements to homes
- poor record keeping
- the illegal use of restraint (physical and through administering excessive medication)
- poor training, that was not considered a priority
- few policies and procedures developed specifically for the supported living services; policies that did exist were not updated, reviewed or monitored
- service users and relatives rarely involved in care planning and reviews
- the Trust's failure to act effectively on referrals about areas of concern.

Investigation into the service for people with learning disabilities provided by Sutton and Merton Primary Care Trust

(Healthcare Commission, 2007)

The Healthcare Commission undertook an investigation of the service provided for people with a learning disability at Sutton and Merton Primary Care Trust following a number of serious incidents in its learning disability service, including allegations of physical and sexual abuse. This investigation included hospital and community home provision. Findings varied across the settings, but areas of concern included:

- the model of care was largely based on the convenience of the service providers rather than the needs of individuals, with a lack of individualised care and insufficient meaningful occupation of time
- provision of activities was very low
- staff lacked experience and training
- there was no policy or system for monitoring restraint, and staff who were using restraint believed that they were not doing so, due to lack of understanding
- there were weaknesses in the implementation of adult protection procedures, such as poor communication, lack of staff awareness about adult protection, and poor follow-up of actions agreed at meeting
- the views of service users were rarely sought and advocacy services were seldom provided
- staff sickness and vacancy levels were high
- supervision and appraisal were either ineffective or not occurring at all
- those care plans that did exist were often not up to date, and there was little evidence of regular reviews of these plans.

The Healthcare Commission concluded that this type of institutional abuse was largely unintentional, but was abuse nevertheless.

REFLECT

Identify the common working practices that have occurred in each of these cases. Select one of the reports and read the recommendations following the investigations. Does your setting act in accordance with each of the relevant recommendations? If not what action do you need to take in order to do so?

Safeguarding procedures and processes

ACTIVITY

Review

- Are you familiar with your organisation's safeguarding procedures?
- Are you familiar with the Local Authority Multi-Agency safeguarding procedures?
- Look at your local authority's web pages regarding safeguarding adults, and identify any material you need to add to your organisation's resources.
- Where do you keep safeguarding procedures, policies and resources and are these accessible to staff? How have you made staff aware of them?
- How do you disseminate new information to staff?
- How do you know that all staff have understood the information?

When you first receive information concerning a service user's well-being, you will need to consider whether it is appropriate to deal with the information as a concern, complaint or an adult abuse alert. The CSCI (2006) defines a **concern** as a general expression of dissatisfaction; a **complaint** as a specific expression of dissatisfaction, and an **alert** as an issue or information of a serious nature indicating possible abusive or criminal practices.

CASE STUDY

Consider each of the following situations and identify whether they are a concern, complaint or alert.

The niece of service user ***Ethel*** *has made an appointment to have a meeting with you. At the meeting, the niece says she is unhappy about the level of cleanliness in her aunt's room, and on a number of occasions has found pieces of old food and dirty underwear on the floor.*

Joyce *has lived in the residential care home you manage for six months. In the past two months, her granddaughter has begun to visit weekly and Joyce is delighted about this. However, after a number of visits staff have informed you that Joyce has appeared confused and has told them she has misplaced some money.*

Sami *spent last week having respite care at your setting. The day after he returned home, his mother calls and informs you that Sami has told her he was given meals with meat in them twice last week and he didn't like it. Sami is a vegetarian.*

You have noticed ***Ella****'s behaviour has changed recently and she has become more moody and aggressive. You ask her key worker to keep a diary to see if you can identify any patterns or triggers to this, and also to speak to Ella during a calmer moment. The key worker reports that Ella's moods and behaviour coincide with Marvin's shifts and when she asked Ella about this, Ella said Marvin had been showing her 'funny pictures' recently and it made her feel 'strange and not nice'.*

Responding to disclosures or alerts of abuse

Receivers of alert and referrals of abuse should respond by:

- remaining calm and not showing shock or disbelief
- listening carefully to what is being said
- not asking detailed or probing questions
- demonstrating a sympathetic approach by:
  - acknowledging regret and concern that what has been reported has happened
  - emphasising that they have done the right thing by sharing the information with you
  - saying that you are treating the information seriously and the abuse is not their fault

Do not:

✘ stop someone who is freely recalling significant events – allow them to share whatever is important to them

✘ ask questions or press the person for more details. This may be done during any subsequent investigation, so it is important to avoid unnecessary stress and repetition for the person concerned. This may also invalidate any evidence if required for a prosecution

✘ promise to keep secrets, or make other promises you are unable to keep

✘ contact the alleged 'perpetrator'

✘ be judgemental – for example, asking 'Why didn't you try and stop them?'

✘ break the confidentiality agreed between the person disclosing the information, yourself and your line manager.

(Adapted from Nottinghamshire Committee for the Protection of Vulnerable Adults, 2007)

- giving the individual information about the steps that will be taken, explaining that you are required to share the information with your line manager, but not with other staff or service users. Your line manager will also need to inform others
- ensuring that any emergency action needed has been taken: for example, medical treatment
- informing them that they will receive feedback as to the result of the concerns they have raised and from whom
- making a written record of what the person has told you.

(Adapted from ADSS, 2005)

Recording information

It may not be appropriate to take notes at the time the disclosure or allegation is being made, but a written report should be made as soon as possible afterwards. You should try to record as accurately as possible what the person said, using their own words and phrases, and record facts and observations. If you do record any opinions, ensure that it is clear that that is what they are. Sign, date and time your report. Detailed records of abuse should not normally be kept on an open file.

Stages of response

The CSCI (2006) provides a helpful flow chart of the stages involved in responding to concerns about a service user's well-being (Figure 7.2). You should also make sure that you are familiar with your local authority's multi-agency procedures, as these may vary slightly and contain additional information, such as resources and contact numbers.

Allegations against staff

Although it is a difficult situation when allegations are made against staff, you must ensure that you act appropriately and quickly. Policies and procedures must be followed in these situations too, and whenever staff are alleged to have perpetrated abuse against a vulnerable adult, the local authority

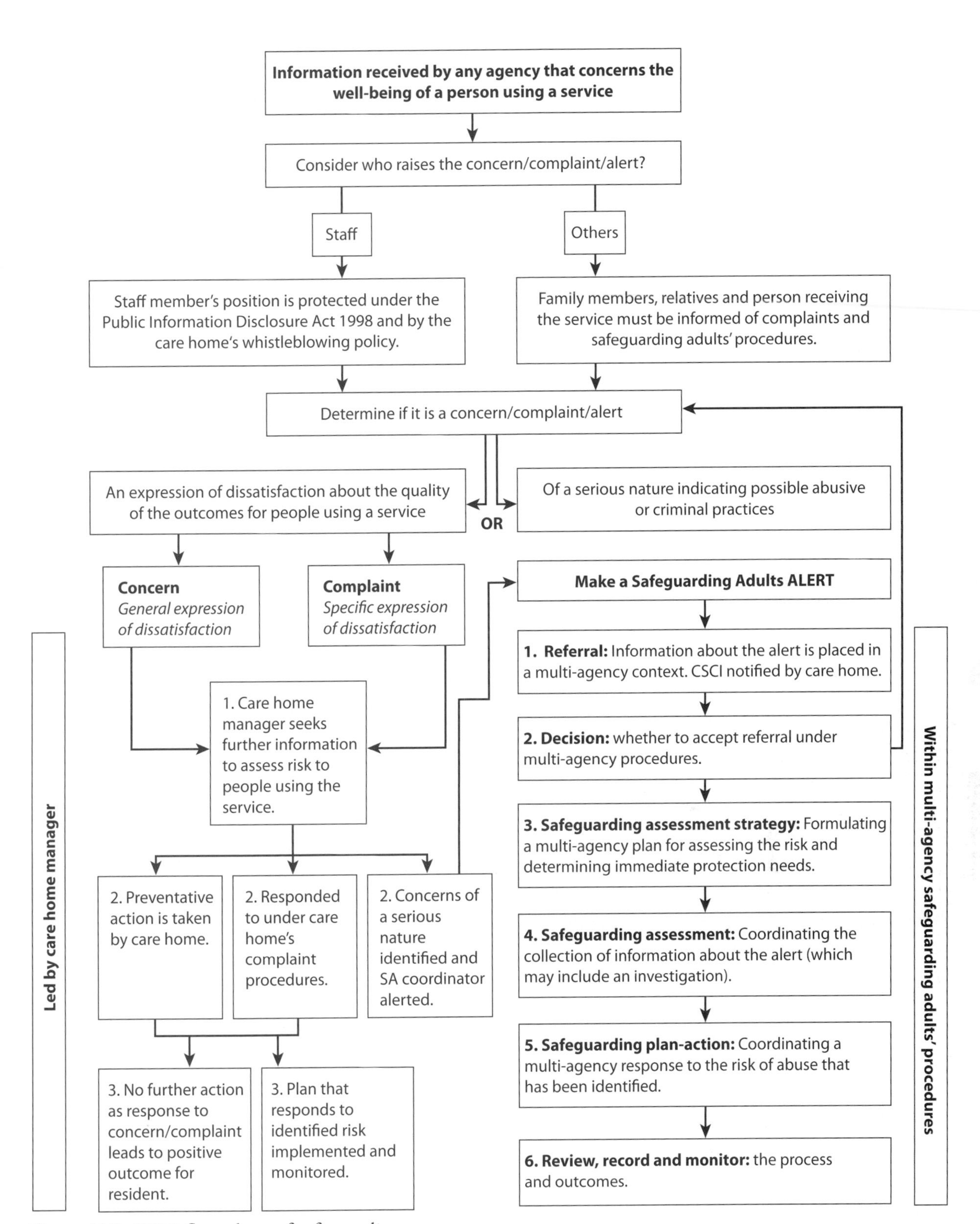

Figure 7.2 CSCI flow chart of safeguarding processes

multi-agency procedures are likely to specify that a Safeguarding Adults Assessment must be undertaken.

You may need to suspend a member of staff while the Safeguarding Adults Assessment takes place, but remember that suspension is a neutral activity – it does not mean the member of staff is guilty. Do not be tempted to transfer staff to other care homes or other services within the organisation rather than suspend them.

Once the assessment is concluded, you will need to decide whether there is a need to instigate capability or disciplinary procedures (see Chapter 5).

Whistleblowing

> 'It was really difficult telling on my colleague who was making old people in our care miserable, but the helpline gave me lots of support. I'm glad I did it.'
>
> (Quotation from Action on Elder Abuse leaflet *You're never too old to hurt*)

The CSCI (2006a) defines whistleblowing as the action whereby 'employees draw attention to bad practice where they work. Employees may blow the whistle within their organisation or, if they do not have confidence in their employer or fear detriment as a result, may disclose their concerns to others.'

National Minimum Standards in all settings refer to the need for registered services to have policies, procedures and/or codes of practice regarding whistleblowing.

Remember: whistleblowers are protected by the Public Interest Disclosure Act 1998.

Whistleblowing is difficult to do and takes courage and determination. The CSCI (2006a) states that service providers that demonstrate good whistleblowing practice:

- show a commitment to high-quality care
- make their staff aware of what is considered good and bad, including abusive, care practice
- encourage frank and open communication about practice from induction onwards
- recognise the value of whistleblowing
- have a whistleblowing policy that sends a clear message that they take incidents of bad practice seriously.

(CSCI, 2006a)

Public Concern at Work (no date) is a national charity promoting whistleblowing and good governance in the workplace. They state that a whistleblowing policy should contain these points

- The organisation takes malpractice seriously, giving examples of the type of concerns to be raised, so distinguishing a whistleblowing concern from a grievance.
- Staff have the option to raise concerns outside of line management and have explained the avenues open to them to raise concerns.
- Staff can get confidential advice from an independent body.
- The organisation will, when requested, respect the confidentiality of a member of staff raising a concern.
- The organisation makes it clear when and how concerns may properly be raised outside the organisation (for example, with the social care regulator).
- It is a disciplinary matter to both victimise a *bona fide* whistleblower and for someone to maliciously make a false allegation.

Review the whistleblowing policy in your setting. Does it contain all of the elements recommended by Public Concern at Work?

Finally, Van den Hende (2001) outlines a series of 'do's and don't's' for whistleblowing (see Figure 7.3).

Taking care of yourself

Working with issues of adult abuse is challenging and can impact on people in different ways. You may experience a range of emotions such as anger, shock, guilt and disgust, and it is important that you take care of yourself. This is easy to lose sight of

Do	Don't
✔ Keep calm ✔ Think about risks and outcomes before you act ✔ Remember you are a witness, not a complainant ✔ Phone Public Concern at Work for advice	✘ Forget there may be an innocent or good explanation ✘ Become a private detective ✘ Use a whistleblowing procedure to pursue a personal grievance ✘ Expect thanks

(Van den Hende, 2001)

Figure 7.3 Dos and donts for whistleblowing

when you are supporting the individual who has been abused, other service users and staff. Make sure that you have support mechanisms to help you address your own feelings and to provide you with the support you need.

Sources of support may be:

- your manager
- colleagues
- human resources personnel or employee assistance schemes or counselling services provided by your employer
- professional organisations such as the British Association of Social Workers (BASW)
- your Trade Union
- your GP.

Although the abuse of vulnerable adults is a challenging and distressing area of work, it is vital that we acknowledge that abuse does take place and that we each have an individual and professional responsibility to take action to prevent it occurring.

> 'Whilst social care services for adults have specific safeguarding responsibilities, everyone in the community has some responsibility. While there are no simple solutions, we have a responsibility collectively to balance people's rights to freedom and to make choices with ensuring that they are safe from harm. We also have a responsibility to ensure that where there are allegations of abuse, the procedures in place are effective and the remedies appropriate.'

(CSCI, 2008:*i*)

NVQ/SVQ

This chapter will help to provide evidence for mandatory units A1 and E1.

It will help you to evidence knowledge statements within the units A2, A3, A4, B2 and B4.

Work Products you may generate and either include in your portfolio or show your assessor to demonstrate your skills and knowledge are:

- **safeguarding policies and procedures that you have written**
- **supervision notes where safeguarding issues have been discussed**
- **relevant training records and plans**
- **minutes of review meetings where safeguarding issues have been raised**
- **risk assessment and management plans that you have written**
- **minutes of meetings, such as network or interprofessional meetings**
- **evidence of your working with advocates.**

References

Action on Elder Abuse (2006) 'What is Elder Abuse?' www.elderabuse.org.uk (accessed 27.03.09)

ADSS (2005) *Safeguarding Adults: A National Framework of Standards for good practice and outcomes in adult protection work*, www.adss.org.uk/publications/guidance/safeguarding.pdf (accessed 12.07.09)

Burgner, T. (1998) *The Independent Longcare Inquiry*, London: DOH

CSCI (2006a) *Guidance: Whistleblowing arrangements in regulated care services*, London: CSCI

CSCI (2007c) *Rights, risks and restraints: An exploration into the use of restraint in the care of older people*, London: CSCI

CSCI (2008) *Raising Voices: Views on Safeguarding Adults*, London: CSCI

DOH and Home Office (2000) *No Secrets: Guidance on developing and implementing multi-agency policies and procedures to protect vulnerable adults from abuse*, London: DOH and Home Office

DOH (2007a) *Independence, choice and risk: a guide to best practice in supported decision making*, London: DOH

Goffman, E. (1963) *Asylums: Essays on the social situation of mental patients and other inmates*, Harmondsworth: Penguin

GSCC (2002) *Code of Practice for Social Care Workers and Code of Practice for Employers of Social Care Workers*, London: GSCC

Healthcare Commission and CSCI (2006) *Joint Investigation into the provision of services for people with learning disabilities at Cornwall Partnership NHS Trust*, www.healthcarecommission.org.uk/_db/_documents/cornwall_investigation_report.pdf (accessed 07.06.08)

Healthcare Commission (2007) *Investigation into the service for people with learning disabilities provided by Sutton and Merton Primary Care Trust* www.healthcarecommission.org.uk/_db/_documents/Sutton_and_Merton_inv_2006_easyread.pdf (accessed 02.05.08)

Lee-Treweek, G. (1998) 'Bedroom Abuse: the Hidden Work in a Nursing Home'. In Allott, M. and Robb, M. (eds) (1998) *Understanding Health and Social Care: An Introductory Reader*, London: Sage

Leicester City Council (2004) *Adult Protection: Preventing Abuse of Vulnerable Adults* (2nd edition)

Nottinghamshire Committee for the Protection of Vulnerable Adults (2007) *Nottingham and Nottinghamshire Safeguarding Adults Multi-Agency Procedure*

Penhale, B., Perkins, N., Pinkney, L., Reid, D., Hussein, S and Manthorpe, J. (2007) *Partnership and regulation in adult protection: The effectiveness of multi-agency working and the regulatory framework in Adult Protection*, London: DOH, Social Care Workforce Unit and The University of Sheffield

Pring, J. (2005) 'Why it took so long to expose the abusive regime at Longcare'

Public Concern at Work (no date) *Best Practice Guide*, London: Public Concern at Work

Scottish Executive (2004) *Investigations into Scottish Borders Council and NHS Borders Services for People with Learning Disabilities: Joint Statement from the Mental Welfare Commission and the Social Work Services Inspectorate*

Van den Hende, R. (2001) 'Public concern at work: supporting public-interest whistle blowing', *Journal of Adult Protection* 3(3):41–44. Cited at SCIE Practice Guide 09: Dignity in care, 2008, www.scie.org.uk/publications/practiceguides/practiceguide09/files/pg09.pdf (accessed 23.09.08)

Wardhaugh and Wilding (1998) 'Towards an explanation of the corruption of care', *Critical Social Policy*, 3: 4–31

Useful reading

ADSS (2005) *Safeguarding Adults: A National Framework of Standards for good practice and outcomes in adult protection work*, London: ADSS www.adss.org.uk/publications/guidance/safeguarding.pdf

CSCI (2006) *Better safe than sorry: Improving the system that safeguards adults living in care homes*. In *Focus: Quality Issues in Social Care*, Issue 5, November 2006

CSCI (2007c) *Rights, risks and restraints: An exploration into the use of restraint in the care of older people*, London: CSCI

CSCI (2008) *Raising Voices: Views on Safeguarding Adults*, London: CSCI

CHAPTER 8

Managing budgets

'A good manager needs to know the advantages of budget setting and management and the perils of not being fully engaged in it.'

(Bamber, 2007)

'We want to remain in control over our own life and money... even in a residential home.'

(CSCI, 2007)

Introduction

Managers of settings have increasing responsibility for budgets within their organisations, and understanding how well your setting is performing financially is important. You will need systems to help you monitor your financial situation, set and manage your budgets and measure the quality and cost-effectiveness of services being delivered. This chapter will focus on the key elements of financial management:

- budget management
- setting budgets
- flexibility in managing budgets
- purchasing
- monitoring budgets
- delegating responsibility
- managing the money of others
- financial controls
- sources of support.

Budget management

The National Minimum Standards Care Homes for Older People Standard 34.1 states that care homes should adopt 'suitable accounting and financial procedures to demonstrate current viability and to ensure there is effective and efficient management of the business'. Similarly, Standard 43.1 in the National Minimum Standards for Care Homes for Adults (18 – 65) refers to the 'financial viability and accountability of the home'.

To manage budgets, you need to be clear what the different types of budget are and what they are for. You may be managing budgets to cover staffing costs, food, utility bills, repairs, activities and petty cash, as well as having an overview of service users' money. Systems that are easy to follow and operate will make the process less time-consuming and help you to identify any potential financial problems. As a budget holder, you will need to monitor all spending against the approved budgets set. This will help you to:

- ensure that budgets are being used effectively
- ensure there are sufficient funds within budgets to meet needs
- demonstrate accountability
- identify under- and overspends and take corrective action where necessary.

REFLECT

- What is your role in the process of budget management?
- What systems do you currently have in place to manage your budgets?
- Who else is involved in managing budgets? What are their roles?

As the budget holder for your setting you should be clear about your role and responsibilities both for the daily management of the finances of your home and the action to take if you identify any difficulties. The Audit Commission, Social Services Inspectorate and The National Assembly for Wales (2004), jointly identify the role and accountabilities of budget holders as follows;

- Budget holders are responsible for contributing to the budget-setting process by inputting information about service trends and activity levels upon which the budget can be calculated, and reaching agreement with finance support staff that the budget is a fair reflection of the cost of these activity levels.
- Budget holders are responsible for responding to budget reports and taking the necessary action to tackle projected budget variances (comparing income/revenue and expenditure with the actual allocated budget). This may include moving resources between budgets to respond to the variations identified in the monitoring processes, in line with the organisation's rules regarding virement (a financial term meaning the transfer of one budget to another). This could be because there is an underspend in one area and an overspend in another. Budget holders are responsible for producing an action plan to deal with the underlying problems.
- If budget holders are not in a position to respond to variances identified in monitoring reports, they are responsible for reporting the issues to a more senior manager. If the variances are such that management action will be unable to address the problem before the year end, this needs to be reported.
- Budget holders are responsible for working with finance support staff to analyse data by inputting their knowledge of the area of service and recent developments in the service.

REFLECT

- Which of these responsibilities do you have?
- What are your responsibilities in relation to variances and virement?
- At what point do you need to report over- or underspends to your manager or finance department?

Bamber (2007) suggests that, to be successful at managing budgets, there are certain things you need to do.

- **Plan ahead and be flexible to change** Budgets are normally planned ahead for at least one financial year, but there needs to be some flexibility to take account of unexpected changes, such as a rise in heating costs. See the section on setting budgets below.
- **Control your costs and manage your cash** You need to monitor and control costs to ensure you are achieving the best value for money. Each element of expenditure should be identified within your costs: stationary, postage, telephone, and so on. Bamber (2007) stresses that 'if you can't measure it, you can't manage it'. It is also important to monitor the spread of your expenditure so that your budget is not spent within three months when it should last for six!
- **Set and measure your key performance targets** Identify some key performance targets to help you to assess the business's financial performance.

These might include areas such as agency worker costs and occupancy levels. The areas you select will help you to identify quickly if there is likely to be an overspend or drop in income.

- **Maximise your income** Review your performance over the past year and identify any areas where you could improve. Maximising income is not just about being paid more; it is about being efficient and effective in budget management.

A number of these areas are outlined in more detail in the sections that follow.

ACTIVITY

Review your budgeting

Give an example of how you are putting each of the areas identified by Bamber into practice. Are there areas of your practice or the systems you have in place that you feel you could develop further? Say how you can improve the area you have identified and devise an action plan to show how you aim to do this.

Manage within your budget

A crucial aspect of managing budgets is to manage **within** your budget, balancing income with expenditure. This means:

- ensuring service plans, commitments and activity levels match the budget available. If they do not, you will need to either increase the resources available or manage the overcommitment by reducing services or costs. You will need to consider how feasible either of these options is
- do not commit yourself to 'efficiency savings' in order to balance the books unless you can specifically identify where the savings will come from.

(Adapted from Audit Commission, SSI and National Assembly for Wales, 2004)

Why might overspends and underspends occur?

During the financial year overspends or underspends might occur for the following reasons.

- Expenditure has been coded inaccurately, i.e. it has been coded against a different budget.
- There has been more activity (and expenditure) than forecast and planned.
- Poor planning and initial budget setting.
- Failure to take changes into account, e.g. the cost incurred as a result of new regulations, or if funding has been moved from one part of the budget to another.
- Delays in processing finances, e.g. invoices.
- Poor management of resources including staff.

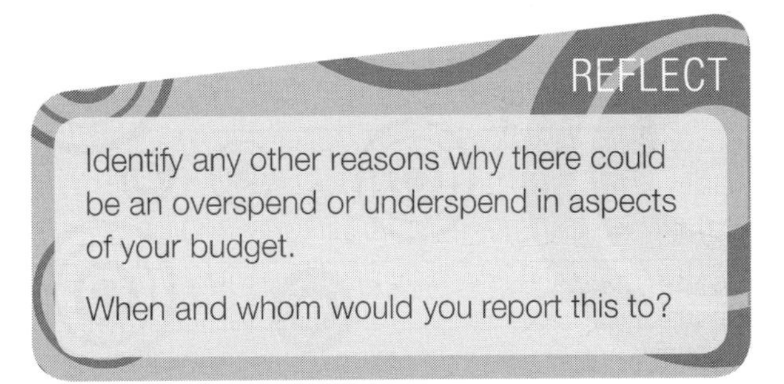

Setting Budgets

To meet the criteria of the NVQ/SVQ Finance Unit, you will need to be involved in the negotiations for the allocation of budgets within your setting. Before setting a budget the Audit Commission, SSI and National Assembly for Wales (2004) identify three key stages which you must go through. These are outlined below.

1. Review the past

This involves:

- monitoring recent trends in demand and expenditure
- monitoring trends in funding streams (for example levels of income from charging)
- monitoring and reporting on actual performance and outcomes, including end-of-year position

CASE STUDY

Reviewing the past

Roger has managed Lantern House residential care home for older people since February 2008. One of his first major tasks was to oversee the opening of Lantern House Annexe in April 2008 which allowed Lantern House to increase its provision of respite care.

Roger's manager, Sarah, has asked him to become involved in reviewing and setting budgets for the next financial year (08–09). The homes financial year runs from July to June each year.

Firstly, Sarah asks Roger to review the previous year (07–08) and particularly focus on costs for agency staff; office equipment; gas and electricity and marketing. All of these varied significantly from the projected budget for the year.

- *Look at the Figures 8.1 8.2 and 8.3 try to identify whether there were particular points in the year when costs peaked or dipped in each category.*
- *What possible reasons are there for these variations from projected costs?*

Cost Area	Projected Costs (£)	Actual Costs (£)
Agency Staff	2000.00	3837.01
Marketing	700.00	00.00
Gas/Electricity	3200.00	5884.85
Office equipment	600.00	825.00
Total	**6500.00**	**10546.86**

Figure 8.1 Projected and actual costs 07–08 in four key areas

Date		Debit	Credit	Balance
1.04.08	Top-up petty cash balance to float of £100: Cash		63.29	100.00
2.04.08	Stationary – Robinsons stationers (paper & printer toner)	43.20		56.80
15.04.08	Stamps	12.35		44.45
17.04.08	Bus fares for J.B. and Margaret W – to accompany J.B. shopping	3.80		40.65
26.04.08	Stationary – Robinsons stationers (paper)	21.87		18.78
	Cash		81.22	100.00

Figure 8.2 Petty cash account April '08

	July 07	Aug 07	Sept 07	Oct 07	Nov 07	Dec 07	Jan 08	Feb 08	Mar 08	Apr 08	May 08	Jun 08	Total
Income	12635.97	12962.54	12635.97	12635.97	12891.36	13103.78	12998.67	12635.97	12962.54	13465.28	13721.87	14123.98	156773.90
Costs:													
Agency Staff	0.00	626.75	231.19	0.00	0.00	783.54	173.64	363.97	0.00	431.19	532.76	693.97	3837.01
Marketing	0.00	0.00	0.00	0.00	0.00	0.00	0.00	0.00	0.00	0.00	0.00	0.00	0.00
Gas/Elec			973.21			1431.56			1962.71			1517.37	5884.85
Office Equip	50.00	50.00	50.00	50.00	50.00	50.00	50.00	50.00	150.00	50.00	150.00	75.00	825.00

Figure 8.3 Monthy account and income in four key areas of expenditure

- identify any outstanding debtors or creditors as these will need to be taken into account in the next years accounts
- considering the outcome of any reports from external inspection and auditors.

2. Forecast the future

This involves:

- considering the impact of national policies and regulations
- identifying and estimating the levels of the various funding streams, e.g. what income do you expect from long-term care and respite care services?
- identifying the length of time a known financial commitment is expected to last (involving expenditure and income), e.g. how long are you committed to contracts with a maintenance company? Is there a time limit on the local authority funding of a placement?
- forecasting the future impact of known trends in demand and expenditure
- forecasting the impact of demographic trends.

3. Set strategies and plans

This involves:

- taking the organisation's business and strategic plans into account, as well as human resource and asset management plans
- identify the knowledge and skills required for effective budget management within your setting, e.g. do you or other staff need additional training?
- be clear about the arrangements for commissioning and procurement of services, e.g. which, if any, aspects of these areas are your responsibility?
- involve others (e.g. other staff) in the financial planning process.

CASE STUDY

Forecasting the future

Having reviewed the past year Sarah asks Roger to look at the same areas (agency staff; office equipment; gas and electricity and marketing) and identify key points for consideration when setting budgets for the following year (09–10). As part of this Sarah asks Roger to identify and review the condition of all office equipment, identifying any which is a priority for replacement in the next year. Sarah also asks Roger to review petty cash expenditure which seemed to increase later in the year.

- *List key questions and areas which you think Roger should consider in relation to the four areas.*
- *Devise a proforma which Roger can use to list all office equipment and justify his priorities for replacement for next year.*

Once these steps have been taken then the final stage is setting budgets.

Scragg (2001) identifies a range of questions to be taken into account when setting budgets.

CASE STUDY

Setting strategies

Roger has spent considerable time reviewing budgets and considering plans for the next financial year. Most of his discussions have taken place with his manager Sarah.

- *How might Roger have involved staff more fully in the financial reviewing and planning processes?*

ACTIVITY

Factors affecting your budgets

Identify the different factors you need to consider when you are setting budgets.

- Who do you involve in this process and why do you involve them?
- Do you have contingency plans?
- If yes, why do you have them, and what are they? If not, why not?

- Is the current budget sufficient for the coming year's activity?
- Is it logical – does it meet the requirements of the different aspects of your service?
- Are there known or likely service changes next year?
- Are there service priorities which you have to take into account?

Standard 43.2 of the National Minimum Standards for Care Homes for Adults (18–65) requires settings to have business and financial plans in place that are available to be inspected annually. To be effective, these need to identify contingency plans to show that you have thought through potential difficulties and will be in a position to deal with them.

Monitoring budgets

It is essential to monitor budgets on a regular basis; this should be carried out a minimum of once a month. This will help you to clearly identify expenditure to date and also if there are any changes in projected spending. Monitoring your budgets regularly will also help you to identify if budgets are being misused.

The types of information needed to monitor budgets are as follows:

- what the budget is for each area of the setting you manage
- all incoming money to date
- all expenditure to date

CASE STUDY

Setting the budget

You are working with your area manager to set the budgets for the coming financial year. You have collected your financial records from the previous year, which outline all income and expenditure to date. You have a maintenance budget, which has been little used during the last two years. As a result, you have transferred the allocated funds to the utilities budget, where there has been a significant rise in costs, and to the service user activities budget. You have recently had a health and safety inspection on the home and it has been highlighted that repairs will need to be carried out in the kitchen and one of the bathrooms.

- *How will this affect the way you allocate budgets for the coming year?*
- *How will you ensure there are sufficient funds to pay for the repairs and the increased utility bills?*
- *Service users have become used to the increase in their activities budget. What implications might the current situation have for service users and staff? If you are going to make changes how will you inform both groups?*

- any variations to spending
- balance of budgets to date
- future spending commitments.

The Audit Commission (2000) remind budget holders that 'good budgeting means not running into deficit, but equally it means not carrying large balances of unspent money from year to year without good reason. This means considering how the budget for a particular area of spending will balance out over a longer period than the next financial year. Any surpluses beyond a small contingency should accrue for a specific purpose.' (Audit Commission, 2000:p6).

ACTIVITY

Monitoring your budget

- How do you keep financial records in your setting?
- Are they up to date?
- How do they meet the financial regulations of your organisation?
- If another manager or staff member had to access them, would it be clear where and how you have recorded the information?
- Are there any areas in which you need to improve your record keeping?

In order to achieve this you will need to monitor budgets effectively, and this means that your records obviously need to be kept clear and kept up to date.

As well as monitoring budgets, you are likely to be required to report to others regarding your budget. You may be required to do so on a monthly basis, but should certainly conduct a quarterly review of your actual budget against the planned budget with the relevant managers in your organisation.

Flexibility in Managing Budgets

You are likely to have some flexibility in how you manage your budget on a day-to-day basis, but you should be clear about the limitations of this. The Audit Commission, SSI and National Assembly for Wales (2004) also stress that there should be clear organisational guidance in relation to virements (the movement of money from one budget to another) which needs to include:

- the cash value for which virements are permitted. These may vary for people holding different levels of responsibility within the organisation
- whether virements can be actioned across different service areas. For example, are budget holders free to vire money from budgets for the maintenance of the building to meet additional staff costs?
- whether the virements allowed are short term or long term, i.e. whether they will cross financial years
- reporting arrangements to senior managers and directors. This enables senior managers and directors to monitor the budget virements which have been actioned and to evaluate their impact.

Purchasing

It is crucial to try to achieve the best value for money possible from all purchases of goods or services. In this context, value for money is about getting the right quality at the best available price. This often means looking further ahead than the immediate purchase, especially when selecting equipment. You must always ensure you take associated costs, such as supplies and maintenance, into account. Organisations often have purchasing procedures in order to prevent waste and fraud. These should specify who is designated to authorise purchases, and to what financial value.

REFLECT

Which goods and services do you have responsibility for purchasing? Remember to include even relatively small areas of purchasing, e.g. stationary, stamps etc.

How often do you review suppliers of goods and services to ensure you are achieving best value for money? How do you complete the review? Who else is, or should be, involved?

Financial Controls

Your organisation may have several systems for processing and recording financial transactions. These may extend from petty cash to purchasing, through payroll and contracting services to income collection systems. The systems are important because they serve to protect the organisation's resources from loss or fraud and they provide information to staff, managers and directors about how the budget is being spent (Audit Commission, 2004).

The Audit Commission (2000) provides the following examples of internal controls:

- internal checks – one person checking another person's work
- separation of duties – distributing the work so that key tasks are assigned to separate members of staff
- systems manuals – clear, readable descriptions of how systems work and who does what
- a system of authorisations – each transaction is authorised before passing on to the next stage of the process
- an audit trail – this tracks all stages of a transaction, for example from copy order to invoice, to accounts, to cheque, and in reverse.

These controls ensure that:

- payments are made only to bona-fide employers and suppliers
- payments are made only for goods and services actually received
- cash transactions, whether income or petty cash expenditure, are secure in all respects.

(Audit Commission, 2000:12)

REFLECT

What internal financial controls are present in your organisation and setting?

Would any of the examples provided by the Audit Commission be helpful to introduce? Why?

Delegating responsibility

As the budgetholder, you have the authority to spend money and to delegate responsibility to staff

CASE STUDY

Delegating financial duties

You have delegated responsibility for food shopping to one of the experienced staff in your team who has said they would like to take on more responsibility. This staff member said they had some experience of this area in a previous job. You did not have time to go through the systems in your setting before he took over this duty, due to staffing shortages. After one month, you check the budget and notice there has been an overspend on the food budget. You have also noticed that food is being thrown away because it has not been used before its sell-by date.

- *Is it the staff member's responsibility that there has been an overspend?*
- *Do you think you prepared the staff member adequately?*
- *What action can be taken to minimise waste and maximise the budget?*
- *How can you support the staff member to develop his skills in this area?*

within your team. This can be beneficial for staff as they can develop a more realistic view of budgets and the limitations that are placed on settings. It can also help staff members who want to develop their skills and knowledge in other areas and take on more responsibility. See the section on delegation in Chapter 5.

Managing the money of others

'People who use social care services should be able to use and manage their money as and when they choose' (CSCI, 2007d).

Figure 8.4 It is important to include service users in the management of their money

As the manager of a residential setting, you will have an overview of service users' money. This will include monitoring how much they have and where, and how it is spent. When managing the money of service users, it is important that they are involved in the decision-making processes regarding how their money is spent. To do this effectively, people need to be kept well informed of the state of their finances.

Some care homes keep separate tins or envelopes containing residents' cash. However, the CSCI is aware of cases where this system has led to significant problems and errors.

- ✘ Staff lending an individual's money to other residents without the knowledge or consent of the resident or the authorised person acting on their behalf.
- ✘ Managers borrowing residents' money for petty cash for the care home, again without consent.
- ✘ An incorrect balance suggesting that money is lost, stolen or borrowed without being replaced.
- ✘ The use of one tin for three or four people because of lack of space, which leads to problems with the accounting system.

Care homes can avoid these problems by establishing an account that separates incomings and outgoings for individual residents and that includes a facility for calculating individual interest

(CSCI, 2007d)

- ✔ Support people to manage their finances independently.
- ✔ Make support arrangements clear, safe and accountable.
- ✔ Make all transactions transparent.
- ✔ Balance accessibility and security.
- ✔ Make procedures more than just a piece of paper.
- ✔ Practice sound financial management.

(CSCI, 2007)

ACTIVITY

Putting it into practice

Using the table below, identify how you put these areas into practice in your setting.

	Practice area	**How do you put this into practice? Give specific examples.**
1.	Are service users consulted about how they want their money spent?	
2.	Are assessments carried out to ascertain if a service user is able to manage their own finances?	
3.	Does the organisation have a named corporate appointee? (A corporate appointee is a registered person recognised by the Department for Work and Pensions to have the authority to manage social security benefits on behalf of people who lack capacity.)	
4.	Are all financial transactions for service users recorded and are all receipts kept?	
5.	Do service users have somewhere secure to store their money in the setting? Do they have easy access to their money?	
6.	Do you have a clear limit about the amount of cash belonging to service users held on the premises?	
7.	Are procedures for managing service users' money clearly written down in a format that everyone can follow and understand?	
8.	Do you keep money and records separate for each service user?	
9.	Is the account, or accounts, used to manage service users' finances separate from the organisation's main business account?	

ACTIVITY

Sources of support

Identify all the sources of support you have in relation to budget management and add them to the image below.

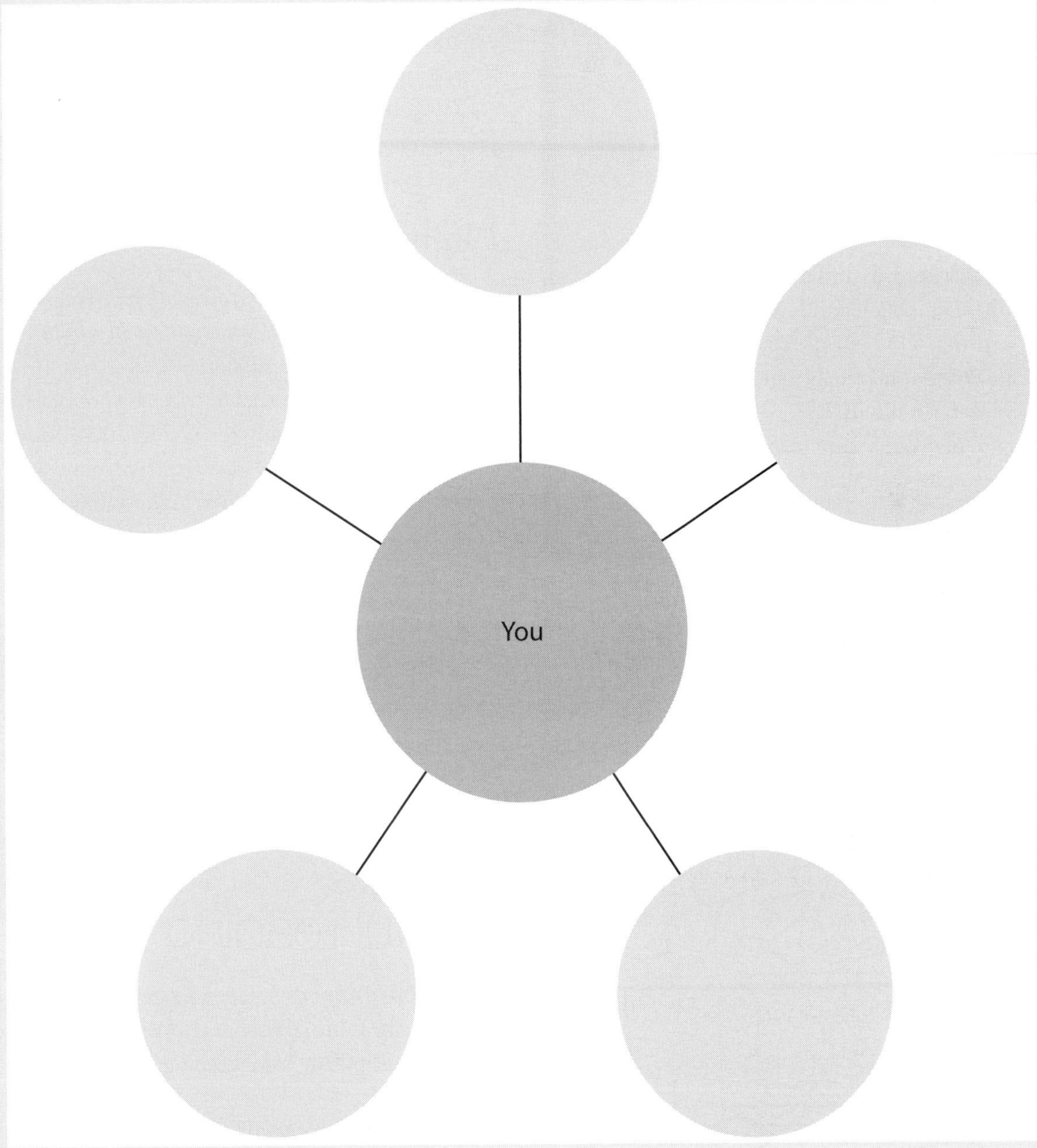

You

Figure 8.5 Sources of support for budget management

The CSCI identified a number of lessons from their inspections of care homes and has devised a good practice checklist to assist care settings when supporting service users to manage their money.

NVQ/SVQ

This chapter will help to provide evidence for Optional Unit E6.

Work Products you may generate and either include in your portfolio or show your assessor to demonstrate your skills and knowledge are:

- **budget records and print-outs**
- **financial policies and procedures you have written**
- **minutes of meetings where financial issues have been discussed**
- **care plans or action plans outlining strategies and plans for supporting service users to manage their finances.**

Figure 8.6 Recognise when to ask for help

Sources of support

Managing budgets can be daunting and confusing, particularly at first. It may be an area you feel is alien to your training and experience as a social care professional. You will become more confident with experience, but it is also important to recognise the sources of support you can access in this area.

Remember, asking for help does not make you a poor manager!

References

Audit Commission (2000) *Keeping your balance: standards for financial management in schools.* London Audit Commission

Audit Commission, Social Services Inspectorate and The National Assembly for Wales (2004) *Making Ends Meet – Financial Management* from www.joint-reviews.gov.uk/money/Financialmgt (accessed 8.7.08)

Bamber, L. (2007) 'Setting Budgets', *Care Management Matters*, April 2007

CSCI (2007d) *In safe keeping: Supporting people who use regulated care services with their finances*, www.csci.org.uk/pdf/in_safe_keeping.pdf (accessed 07.08.08)

Scragg, T. (2001) *Managing at the Front Line: A handbook for managers in social care agencies*, Brighton: Pavilion

Useful reading

CSCI (2007d) *In safe keeping: Supporting people who use regulated care services with their finances*, www.csci.org.uk/pdf/in_safe_keeping.pdf (accessed 04.10.08)

ARC (2005) *My Money Matters! Guidance on best practice in handling the money of people with a learning disability*, www.arcuk.org.uk/ (accessed 18.12.07)

CHAPTER 9

Interprofessional working

'I have social workers, occupational therapists, district nurses and others and they never talk to each other. Every time a different worker comes, they ask me what's wrong with me. I say, 'Haven't the other people told you?' and they say, 'We don't do that.' A friend suggested I throw a Christmas party for them all, get them all here and make them talk to each other.'

(Beresford 2005:11)

Introduction

This chapter focuses on the need for, and barriers to, interprofessional working. The chapter highlights the drive to improve interprofessional working and the benefits this can bring for service users, carers and the professionals involved.

The chapter covers:

- terminology
- the importance of interprofessional working
- factors that help or hinder interprofessional working.

Terminology

The literature on interprofessional working makes use of a wide range of alternative terms – including 'inter-agency', 'multi-professional', 'multi-agency' and 'multi-disciplinary' – but for the purposes of this publication, the term 'interprofessional' will be used. This supports the explanation provided by Barrett et al. (2005), who view the use of the prefix 'inter' as suggestive of collaboration, and interprofessional working as 'the process whereby members of different professions and/or agencies work together to provide integrated health and/or social care for the benefit of service users'. This final point – 'for the benefit of service users' – is the most central aspect of interprofessional working, and one that can easily be lost in the context of competing demands, increasing pressures and diminishing resources.

The importance of interprofessional working

The idea, and practice, of interprofessional working in health and social care is certainly not new, but has had considerable and consistent emphasis from the government over the last decade. Its importance is stressed in a number of contexts.

Legal and policy context

A significant amount of legislation and policies require or emphasise the need for interprofessional working. A number of these are listed here.

- The Health Act 1999 enabled money to be pooled between health bodies and health-related local authority services, and also allowed for the integration of resources and management structures.
- The Single Assessment Process introduced in the National Service Framework for Older People (DOH, 2001a) emphasised the need for agencies to work together to ensure that assessment and care planning are effective and coordinated.
- The Green Paper *Independence, Well-being and Choice in Adult Social Care* (DOH, 2005) which outlines the government's future vision for social care, identifies the need to strengthen joint working between health and social care services as one of the strategies required to enable them to achieve the outcomes identified within the Green Paper.
- Objective 11 of *Valuing People* (DOH, 2001b) is 'to promote holistic services for people with learning disabilities through effective partnership working between all relevant local agencies in the commissioning and delivery of services'.
- As indicated by its title, the whole focus of *No Secrets: Guidance on developing and implementing multi-agency policies and procedures to protect vulnerable adults from abuse* (DOH, 2000) is on interprofessional working for the prevention of, and the protection of, vulnerable adults from abuse.

This list is by no means exhaustive. You should consider which other legislation and policies that you work with also emphasise interprofessional working.

Lessons from inquiries and serious case reviews

Chapter 8 includes outlines of some of the recent inquiries and serious case reviews, and highlights key messages from these reports, including the consequences of the failure of interprofessional working, and particularly poor communication (Reder et al., 1993). Indeed, O'Rourke (1999) highlighted that more than 35 inquiry reports had been published since the Christopher Clunis Report (1994), each making similar recommendations for improvements in interprofessional working. The Victoria Climbié inquiry (Laming, 2003) has been the most recent high-profile inquiry widely publicising the failure of interprofessional working, and the repetition of the same mistakes.

The Serious Case Review into the murder of Steven Hoskin (Flynn, 2007) also identified significant failings in this area: 'What is striking about the responses of services to Steven's circumstances is that each agency focused on single issues within their own sectional remits and did not make the connections deemed necessary for the protection of vulnerable adults and proposed by *No Secrets* (Home Office/Department of Health, 2000)'.

These inquiries and serious case reviews serve to remind us of some of the most serious consequences of failing to work interprofessionally.

Improved outcomes for service users and carers

Although research in this area is still relatively limited, it does highlight a range of benefits for service users and carers from interprofessional working. According to Dawson and Barlett (1996) and Atkinson et al. (2002), these include:

- easier and/or quicker access to services
- early identification and intervention. Research indicates that service users rate 'low-level' preventative services highly because of the impact on their quality of life (Moriarty, 2005)
- better quality services
- the reduced need for more specialist services.

What service users and carers want

Unsurprisingly, given the benefits identified above, service users and carers are clear that they want the professions to work together. Service users and carers have indicated that they prefer a single point of contact for services, and a named person to coordinate the involvement and work of professionals, as well as information sharing between professionals, and by professionals with them (service users and carers) (Mukherjee et al., 1999).

Benefits for staff

Interprofessional working has also been shown to have benefits for staff. Dawson and Barlett (1996) found that there was more clarity regarding roles and responsibilities, while Sloper (2004) identified the links between interprofessional working and lower levels of stress for staff.

Factors that help or hinder interprofessional working

The literature in this area cites a broad range of factors that help or hinder interprofessional working, often dividing them into different categories (Means et al., 2003; Barrett and Keeping, 2005; Quinney, 2006). For simplicity, some of these factors have been combined below within three main categories,

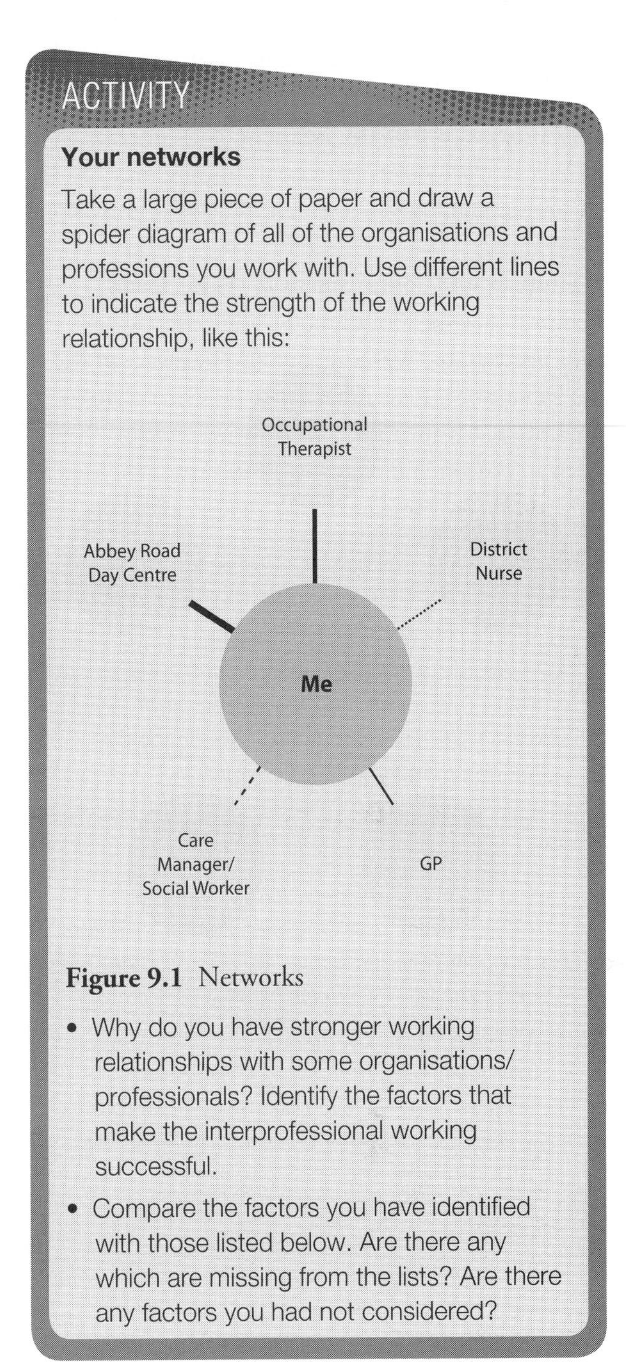
ACTIVITY

Your networks

Take a large piece of paper and draw a spider diagram of all of the organisations and professions you work with. Use different lines to indicate the strength of the working relationship, like this:

Figure 9.1 Networks

- Why do you have stronger working relationships with some organisations/ professionals? Identify the factors that make the interprofessional working successful.
- Compare the factors you have identified with those listed below. Are there any which are missing from the lists? Are there any factors you had not considered?

although several factors could apply within more than one category.

Structural and procedural (including financial)

Policy and procedure differences, or lack of procedures/guidance. Where these elements exist, it will naturally be more difficult for professionals to work collaboratively, and conflict and/or uncertainty between professionals is likely to increase.

Resources, especially finances. Lack of resources or lack of clarity over sources of funding may result in conflict, and may also result in delay of provision of services.

Support and commitment at senior levels. Senior managers should model good practice in interprofessional working, but also be aware of the anxiety and resistance that can arise when changes are imposed from above. Hornby and Atkins (2000, cited in Barrett and Keeping, 2005) stress the importance of enabling professionals to contribute to discussion and decision making. This is likely to lessen anxiety and uncertainty about roles, responsibilities and the changes ahead.

Competing or conflicting targets. Performance management targets and indicators have become more prominent in the government's drive to improve and measure the quality of health and social care services. Where focus on targets takes precedence over other aspects of service delivery, or where targets are competing or conflicting between

ACTIVITY

Who Holds the Power?

Think about a situation where interprofessional working occurs in your workplace, such as a service user's review. Answer the following questions in relation to that situation:

1. Who defines the issues, agenda items or 'problems'?
   For example, does everyone make a contribution to setting the agenda or is this specified by one professional? Who decides the order items are discussed in?
2. Whose terms are used?
   Does everyone involved communicate through the use of commonly understood terms or does one particular professional language predominate. For example, is there a heavy reliance on medical terminology?
3. Who controls the domain/territory?
   Is control mutually negotiated, does it vary in accordance with which professional's knowledge and expertise best fits the particular needs of the service user at a given point in time, or does control sit predominantly with one professional group? For example, do the majority of those involved look to one professional (or professional group) to lead; does one professional group always chair the reviews?
4. Who decides on what resources are needed and how they are allocated?
   Are resource issues mutually agreed or dictated by one professional group? Who holds power over resource allocation?
5. Who is responsible?
   Is everyone's responsibility recognised or is it assumed that one professional group will have overall responsibility? Does one professional group dictate the responsibilities of others?
6. Who prescribes the activity of others?
   Is there joint agreement regarding the activities of those involved or does one professional group prescribe activities? For example, does the social worker decide tasks, targets and timescales following a review or are these agreed jointly?

(Adapted from: Barrett, G and Keeping, C, 2005)

- Identify which professional group(s) was most powerful in the situation you reflected on.
- Why do you think this was? Did they hold a particular power, for example over resources or knowledge?
- How did this make you feel and act?
- How do you think this made the service user (and carers) feel and act?
- Having reflected on the issue of power is there anything you would do differently in this situation?
- What could you do to minimise the power imbalance between service users (and carers) and professionals?

services, the environment for interprofessional working is made more challenging.

Power. It is beneficial to share power between professionals (as well as service users and carers), but there need to be clear levels and lines of responsibility and accountability. Without them, it is easier for professionals to negate their responsibility and for a 'blame culture' to develop.

Issues of power can arise in a range of contexts in interprofessional working, not just in relation to structures and procedures. For example, areas such as professional culture and communication, which are outlined in the following sections, contain aspects of power. Use the activity on the previous page to reflect on issues of power in more depth.

Professional (including cultural)

Knowledge and understanding of professional roles. It is vital that you understand and are clear about the roles and responsibilities of other professionals. It is also important that you are clear about your own role and responsibilities and can communicate this to others.

Confidence. Both personal and professional confidence are important and link with clarity of role and responsibility. Molyneux (2001, cited in Barrett and Keeping, 2005) found that those 'professionals who were confident in their role were better able to work flexibly across professional boundaries without feeling jealous or threatened'.

Envy. Envy occurs particularly over issues of resources and power. It is likely to be heightened where resources and professional power are limited.

Professional culture and stereotypes. There are cultural differences between professional groups, which may stem from differences in training, values, perceived status, differences in professional language (or jargon!), roles, and perspectives on need. These elements may also influence the stereotypes which may be held by one professional group about another.

A key cultural difference is that resulting from the allegiance to either the medical or the social model of care (see Chapter 3 for definitions of these models). This affects the way in which need or 'problems' are perceived and approached, and how work with service users and carers is undertaken.

Trust and mutual respect. Both personal and professional trust and mutual respect are factors in interprofessional working. It is unsurprising that research confirms that interprofessional working is likely to be successful where there is trust between those involved (Rummery, 2002, cited in Means et al., 2003).

Personal

Willing participation. Individuals must be willing to participate in interprofessional working, and to see this as a positive way of working. Without this approach and level of personal motivation, whatever structures and procedures are in place to promote interprofessional working, it is less likely to succeed.

Open and honest communication. This is a key element at all levels of interprofessional working. It means professionals need to be willing to both listen actively and to provide constructive feedback.

Defences against anxiety. Working within health and social care services can be a stressful and anxiety-provoking activity. Individuals may displace their anxieties onto other professionals as a reaction to their situation and as a coping mechanism.

Conflict. Conflict is often seen as negative, but it is important to remember that conflict can be positive and is often needed in order to bring about change.

Length and quality of working relationship. However difficult the context of health and social care work, positive, trusting relationships built over time between two or more professionals can overcome many of the challenges, and result in positive outcomes for service users and carers. The value of taking time and effort to build relationships should not be underestimated.

REFLECT

Which of the factors could you draw on to help improve the interprofessional relationships you identified as weaker or more challenging in the activity 'Your networks'?

In conclusion, however difficult it may be, all professionals have a responsibility to find the time and strategies to make interprofessional working a reality. The importance of interprofessional working for service users and carers is stressed again in the following comment from an Age Concern Consultation Exercise (2004):

> 'It seemed like quite a few people had pieces of the jigsaw, but no-one had the picture on the box.'

NVQ/SVQ

This chapter will help you to provide evidence for the following units:

- **mandatory units E1**
- **optional units D1, D2, D3**
- **the following elements from Units A1: A1.1 and A1.4.**

Work Products you may generate and either include in your portfolio or show your assessor to demonstrate your skills and knowledge are:

- **minutes of any interprofessional meetings**
- **records of care plans and reviews**
- **feedback, review or evaluation of external services which have been used).**

CASE STUDY

Consider these three scenarios.

Scenario one *A review is scheduled for a service user in your unit. You believe it will take the hour and a half set aside for it, as there are some complex issues for discussion. The social worker calls you to ask if you can 'squeeze in' a review for another service user, as she (the social worker) is behind with reviews and is under pressure from her manager to catch up.*

Scenario two *At your request, the GP comes to visit one of the service users in the unit who is recovering from a minor stroke. The GP says she will arrange for the OT to visit, but when you speak to the OT later in the week, she says that she knows nothing about the service user or referral. The OT also says that, from your description of the issues, it sounds like it is a physiotherapist who is needed, not an OT, and that she will not attend.*

Scenario three *You manage a residential unit for people with learning disabilities. Sophie, one of the service users, has been asking for more freedom to go out on her own. The professionals involved all agree that Sophie would be able to do this with an acceptable level of risk and that a plan should be developed to enable this to happen. However, Sophie's parents disagree strongly, saying that Sophie must be accompanied at all times. A meeting is arranged for everyone to attend, including Sophie and her parents, to discuss this further. However, two professionals don't turn up for the meeting and those present don't seem willing to tell Sophie's parents directly that they disagree with their views.*

For each scenario, consider these questions.

- *Why might this be happening?*
- *What strategies can you adopt to deal with the situation while maintaining a positive working relationship with the other professionals?*

Figure 9.2 Interprofessional working – putting the pieces together

References

Atkinson, M., Wilkin, A., Stott, A., Doherty, P. and Kinder, K. (2002) *Multi-agency working: a detailed study*, Slough: NFER

Age Concern Consultation Exercise (2004). Cited in *Better Outcomes for Older People: Framework for Joint Services* (2005) The Scottish Executive, COSLA and NHS Scotland

Barrett, G., Sellman, D. and Thomas, J. (eds) (2005) *Interprofessional Working in Health and Social Care: Professional Perspectives*, Basingstoke: Palgrave Macmillan

Barrett, G. and Keeping, C. (2005), 'The Processes Required for Interprofessional Working'. In Barrett, G., Sellman, D. and Thomas, J. (eds) (2005) *Interprofessional Working in Health and Social Care: Professional Perspectives*, Basingstoke: Palgrave Macmillan, pp18–31

Dawson J. and Barlett E. (1996) 'Change within interdisciplinary teamwork: one unit's experience', *British Journal of Therapy and Rehabilitation*, 3:219–22

DOH (1994) *The Report of the Inquiry into the Care and Treatment of Christopher Clunis*, London, HMSO

DOH and Home Office (2000) *No Secrets: Guidance on developing and implementing multi-agency policies and procedures to protect vulnerable adults from abuse*, London: DOH and Home Office

DOH (2001a) *National Service Framework for Older People*, London: DOH

DOH (2001b) *Valuing People: A New Strategy for Learning Disability for the 21st Century*, London: The Stationary Office

DOH (2005) Social Care Green Paper *Independence, Well-being and Choice in Adult Social Care*, London: DOH

Flynn (2007) *The Serious Case Review into the murder of Steven Hoskin* (Flynn, 2007)
Lord Laming (2003) *The Victoria Climbié Inquiry: Summary Report of an Inquiry*, Cheltenham: HMSO
Means, R., Richards, S and Smith, R (eds) (2003), *Community Care: Policy and Practice* (3rd edition), Basingstoke: Palgrave Macmillan
Molyneux, J. (2001), 'Interprofessional Teamworking: what makes a team work well?', *Journal of Interprofessional Care* 15:29–35. Cited in Barrett, G. and Keeping, C. (2005) *The Processes Required for Effective Interprofessional Working*
Moriarty, J. (2005), 'The future of social care', *Journal of Dementia Care* 13(3):10–11
Mukherjee, S., Beresford, B. and Sloper, T. (1999) *Unlocking Key Working: An Analysis of Keyworker Services for Families with Disabled Children*, Bristol: The Policy Press
O'Rourke, M. (1999) 'Dangerousness: how best to manage the risk', *The Therapist* 6(2)
Quinney, A. (2006) *Collaborative Social Work Practice*, Exeter: Learning Matters
Reder, P., Duncan, S., Gray, M. and Stevenson, O. (1993) *Beyond Blame: Child Abuse Tragedies Revisited*, Abingdon: Routledge
Rummery, K. (2002) 'Disability, Citizenship and Community Care: A Case for Welfare Rights'. Cited in Means, R., Richards, S. and Smith, R. (eds) (2003) *Community Care: Policy and Practice* (3rd edition), Basingstoke: Palgrave Macmillan
Sloper, P. (2004) 'Facilitators and barriers for co-ordinated multi-agency services', *Child Care, Health and Development*, 30(6):571–580

Useful reading

Barrett, G., Sellman, D and Thomas, J (eds) (2005), *Interprofessional Working in Health and Social Care: Professional Perspectives*, Basingstoke: Palgrave Macmillan
Every Child Matters (2007) Multi-agency services: toolkit for managers, www.everychildmatters.gov.uk/deliveringservices/multiagencyworking/managerstoolkit/

CHAPTER 10

Continuing professional development

Introduction

Chapter 3 outlined many benefits in developing the workforce in terms of retention, motivation and effectiveness at work. Here we focus on the benefits of, and strategies for, attending to your own needs for continuing professional development (CPD). We start by providing definitions of CPD and considering its value, then explore a number of strategies for planning and managing CPD. Specific attention will also be given to stress management and time management – two key areas in managing self, and work, effectively.

The chapter covers:

- definitions of CPD
- benefits of CPD
- strategies for planning and managing your own CPD
- working outside your comfort zone
- stress management and well-being
- time management.

Definitions of CPD

The Department of Health defines continuing professional development as

> 'a range of learning activities through which professionals maintain and develop throughout their career to ensure that they retain their capacity to practice safely, effectively and legally within their evolving scope of practice'.

(HPC, 2004:7).

The Chartered Institute for Personnel and Development (CIPD) emphasises the individual responsibility professionals must take for their ongoing development within their definition. They also emphasise that this is an ongoing process, rather than merely individual events or activities. CIPD defines continuing professional development as

> 'a combination of approaches, ideas and techniques that will help you manage your own learning and growth. The focus of CPD is firmly on results – the benefits that professional development can bring you in the real world'.

(CIPD, 2008b:1)

Benefits of CPD

Professionals, organisations and research studies cite a range of benefits of, and motivations for, undertaking CPD. Benefits of CPD are generally identified as some or all of the following:

- helps to ensure that you are offering the best service that you can to service users, carers and families
- helps you to identify areas of development that are relevant and support your career
- reassures employers and the public that individuals are competent to undertake a role/tasks
- helps maintain competence
- helps maintain and increase confidence
- updates skills and knowledge/to develop expertise
- helps to prepare for career progression (promotion and new jobs)
- provides challenge and helps to maintain motivation
- helps you learn and develop from feedback from others (service managers, colleagues, managers and others).

Individual motivations for undertaking CPD will also vary. Rothwell and Arnold's (2005) study identified six motivations for undertaking CPD:

- to avoid losing your licence to practice (or registration with professional body)
- because it is enjoyable
- to make up lost ground
- to maintain your current position
- to get ahead of the competition
- to affirm your identity as a good professional.

CPD has been seen as an expectation of individual professionals, and as good practice, for some time. For example, the Skills for Care Leadership and Management strategy (Skills for Care, 2006:31) states that leaders and managers 'need to undertake planned CPD activities to:

- continually develop competences and new styles of working to bring about the change and cultural shifts required to implement the modernisation agenda
- develop and maintain practice that is self-aware and critically reflective
- take responsibility for the ongoing development of self and others.'

Increasingly, CPD has become a mandatory activity. This is evident through the requirements laid down by professional bodies such as the General Social Care Council (GSCC) and Health Professions Council (HPC), as you can see from these extracts from their regulations:

GSCC

- As a social care worker, you must be accountable for the quality of your work and take responsibility for maintaining and improving your knowledge and skills.
- Post-registration training and learning (PRTL) is a key condition for continued registration.

(GSCC, 2002:6)

NB: 'PRTL' is the phrase the GSCC has chosen to use, and is equivalent to other bodies' use of the term 'CPD'.

HPC

'CPD is an important part of your continuing registration. Our standards now mean that all health professionals must continue to develop their knowledge and skills while they are registered.'

(HPC, 2006:3)

Strategies for planning and managing your own CPD

It is important for you to consider, in conjunction with your employer, how you will plan and manage your CPD, rather than relying solely on your employer to manage this process on your behalf. Figure 10.1 shows the main stages involved.

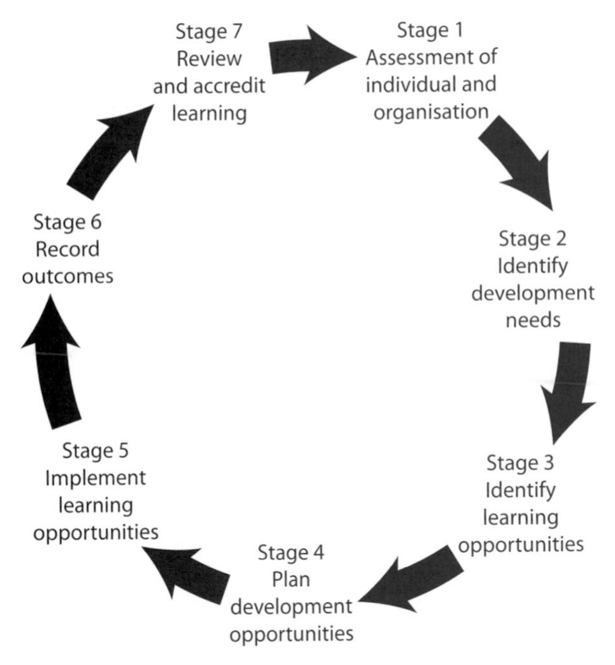

Figure 10.1 Planning and managing your CPD

Stages 1 and 2: Assessment and identification of CPD needs

A range of methods can be used to identify your CPD needs – below are some examples you may wish to follow. You should also review the section on learning needs analysis in Chapter 4 (see page 68).

- **Carry out a Skills/Knowledge Gap Analysis.** First, return to the activity in Chapter 1 (*Reflecting on your job role*, page 7), and list the aspects of your job role you need to develop. Then look at each aspect of your role and rank your skills and knowledge in relation to each area, as shown in Figure 10.2.

Aspect of job role	Knowledge (K) and Skills (S) Ranking (1 is lowest, 10 is highest)
Facilitate the development of staff by arranging duty rotas, delegating appropriate duties and in conjunction with the Workforce Development Manager arrange suitable training and learning opportunities	K 1 2 3 4 5 6 7 8 9 10 S 1 2 3 4 5 6 7 8 9 10
Maintain regular communication with the commissioners of the services to ensure that the provision responds to the assessed needs of each individual resident and provides an early warning of any changes in behaviours that may require an adjustment to the care plan	K 1 2 3 4 5 6 7 8 9 10 S 1 2 3 4 5 6 7 8 9 10
Ensure that suitable and effective strategies are in place and constantly working towards marketing the home and the quality services provided	K 1 2 3 4 5 6 7 8 9 10 S 1 2 3 4 5 6 7 8 9 10
Prepare and present reports and statistical information as and when required	K 1 2 3 4 5 6 7 8 9 10 S 1 2 3 4 5 6 7 8 9 10
In conjunction with the company Finance Officer, ensure that the home operates effectively and efficiently within a specified budget, making the best possible use of available finance	K 1 2 3 4 5 6 7 8 9 10 S 1 2 3 4 5 6 7 8 9 10

Figure 10.2 Skills/Knowledge Gap Analysis example

You should also link this exercise with areas identified in your appraisal with your line manager. Prioritise the three areas you think are most important for you to focus on in the next year.

- **Complete a SWOT analysis**. SWOT is an acronym for Strengths, Weaknesses, Opportunities, Threats (see the descriptions below). Make a list under each section, involving others if this is helpful. Then identify the most important items within the weaknesses and threats sections, and consider your development needs in relation to these areas.
- **Evaluate your current performance against relevant sets of national standards** such as Customer Service Excellence or Investors in People. You could consider whether you are fulfilling your role in meeting these standards or not, and then identify areas for improvement, or undertake a Skills/Knowledge Gap Analysis, as in Figure 10.2 on page 155.

Strengths Identify your strengths: for example time management, team working	**Weaknesses** Identify areas you find difficult or more challenging: for example, managing conflict, budgets
Opportunities These are external factors that will provide opportunities to the organisation and potentially to you: for example, a new senior manager, new funding streams	**Threats** These are external factors which are outside your control (and perhaps the organisation's control) which pose a 'threat' or risk: for example a new provider opening nearby, new regulations which will increase your workload

Figure 10.3 A SWOT analysis

Examples of elements from the Customer Service Excellence Standard

2.2.1 We can demonstrate our commitment to developing and delivering customer-focused services through our recruitment, training and development policies for staff.

2.2.5 We value the contribution our staff make to delivering customer-focused services, and leaders, managers and staff demonstrate these behaviours.

3.4.3 We interact within wider communities and we can demonstrate the ways in which we support those communities.

4.3.3 We give staff training and guidance to handle complaints and to investigate them objectively, and we can demonstrate that we empower staff to put things right.

Stage 3: Identifying CPD opportunities

VIEWPOINT

Identify the various ways that knowledge and practice can be developed, and professional development facilitated. Remember – it is not only formal training that facilitates learning and development: for example, shadowing or reading may be more appropriate than attending training. List as many types of CPD activity as you can think of, and note how many different types of CPD activity you have been involved in within the last three months.

Both the GSCC and the HPC have produced extensive lists highlighting the wide variety of activities that can form part of an individual's continuing professional development.

Arranging to shadow the work of a colleague in a related team or profession
Negotiating protected time to research latest policy and good practice developments in their field of practice
Doing research related to their practice
Completing a period of secondment in another area of related work
Identifying and recording learning gained from individual appraisal
Learning from reflection gained after a particular case or activity
Team meetings or team training events that benefit individual development
Attending meetings or conferences where there are identifiable learning outcomes
Reading an article, report or document where insight or learning results
Doing a new task
Taking on new or different responsibilities
Assessing risk and taking an informed judgment, then reflecting on its outcome
Doing research in relation to writing a report, giving a presentation, leading a discussion or running a seminar or group

(GSCC, no date)

Figure 10.4 GSCC's list of CPD opportunities

Work-based learning	**Professional activities**
• Learning by doing • Case studies • Reflective practice • Audit of patients • Coaching from others • Discussions with colleagues • Peer review • Gaining and learning from experience • Involvement in the wider work of your employer (for example, being a representative on a committee) • Work shadowing • Secondments • Job rotation • Journal club • In-service training • Supervising staff or students • Visiting other departments and reporting back • Expanding your role • Significant analysis of events • Filling in self-assessment questionnaires • Project work • Evidence from learning activities undertaken as part of your progression on the NHS Knowledge and Skills framework	• Involvement in a professional body • Membership of a specialist interest group • Lecturing or teaching • Mentoring • Being an examiner • Being a tutor • Branch meetings • Organising journal clubs or other specialist groups • Maintaining or developing specialist skills (for example, musical skills) • Being an expert witness • Membership of other professional bodies or groups • Giving presentations at conferences • Organising of accredited courses • Supervising research • Being a national assessor • Being promoted *(continued overleaf)*

Formal/educational	Self-directed learning	Other
• Courses • Further education • Research • Attending conferences • Writing articles or papers • Going to seminars • Distance learning • Going on courses accredited by a professional body • Planning or running a course	• Reading journals or articles • Reviewing books or articles • Updating your knowledge through the internet or TV • Keeping a file of your progress	• Public service • Voluntary work

(HPC, 2006)

Figure 10.5 HPC's list of CPD opportunities

ACTIVITY

CPD activities

- Are there any activities you have been involved in within the last three months that you would not have considered as a CPD activity?
- Can you now identify your learning from these activities?

Stages 4 and 5: Planning and implementing CPD opportunities

Once you have identified your professional development needs, you will need to identify the best method of meeting them, and plan how and when you will do so. You should also set a date to review your plan at the outset. On the opposite page there is an example of a CPD action plan.

What? *What is your CPD need?* *What do you need to learn/ achieve/research?*	**Why?** *How does this link with your career plan/appraisal/organisation's business plan?*	**How?** *What type of activity do you need to undertake to achieve the objective?*	**When?** *What is your timescale for completion?*
1.			
2.			
3.			
4.			
Date to be reviewed:			

Figure 10.6 Example of a CPD action plan

Figure 10.7 Reviewing your CPD will allow you to monitor if you are reaching your targets

ACTIVITY

CPD action planning

Imagine that you have completed some activities to identify your CPD needs, and have prioritised the following three areas as your main professional development needs for the next year.

1. to be better at saying 'No'
2. to understand how my team budget works
3. to chair meetings more effectively.

Practice CPD action planning for these three areas using the example format provided.

Stages 6 and 7: Recording and reviewing CPD

It is useful to keep your own record of your CPD activities (this may be a requirement if you are registered with a body such as the GSCC or HPC). On the following two pages there are two examples of records you could use. You may choose to adapt these to suit your own needs or preferred style, or to meet the requirements of your registering organisation. These examples may also be useful as evidence for your NVQ/SVQ.

The Skills for Care Leadership and Management strategy (Skills for Care, 2006:31) states that CPD 'should focus on outcomes and achievements and reflect learning in terms of impact and clear benefits for the user of the service', so remember to also consider the benefits of your CPD activities to service users.

Date	**Duration** (hours/days)	**Type of CPD activity** (reading, shadowing, project work, etc.)	**Details of CPD activity**	**Outline your learning from the activity, implications for your practice and any further action required**

Figure 10.8 Example of a CPD activity record

Name:_______________________________________ Date:____________
Details of CPD activity
Learning gained from the activity
How has this learning been applied to/implemented in practice?
Did you 'cascade' the information to members of your work team? If yes, how and why; if no, why not?
Any future CPD activity or other action required? Identify plan and timescale
Signature:_______________________________________ **Date:**__________

Figure 10.9 CPD activity record and learning log

Working outside your comfort zone

Working inside your comfort zone refers to carrying out your work in ways, and with people, you are familiar with. Part of continuing your professional development often involves working outside your comfort zone: that is, working in new ways and with different people.

REFLECT

- Why do managers need to set new challenges for themselves and others?
- How can this help develop practice?

Being part of a 'learning organisation' involves continuing to challenge yourself and others in the organisation to improve, to learn from mistakes and to try new and different ways of working. This will usually involve you moving outside your comfort zone.

VIEWPOINT

Identify a time when you have worked outside your comfort zone.

- How did you feel and what did you learn from the experience?
- How did this contribute to your professional development?

Although you are likely to have felt anxious and uncomfortable when initially working outside of your comfort zone, by doing so you will have developed new skills and knowledge and increased confidence. You will also have extended your comfort zone and will hopefully feel ready for your next challenge!

Specific attention will now be given to stress management and time management, two key areas in managing yourself, and your work, effectively.

Stress management and well-being

The Health and Safety Executive (HSE) defines stress as 'the adverse reaction a person has to excessive pressure or other types of demand placed upon them' (International Stress Management Association et al., 2004:2). Many things at work can cause stress, and individuals will find some things more stressful than others. It is useful to be aware of the areas of work that you find particularly stressful, as this is the first step towards managing it.

The HSE identifies six key factors that can cause work-related stress:

- the **demands** of your job
- your **control** over your work
- the **support** you receive from managers and colleagues
- your **relationships** at work
- your **role in the organisation**
- **change** and how it is managed.

REFLECT

- Which of the areas listed by the HSE is the most likely to cause you work-related stress?
- Why do you think this is?
- What strategies do you currently have for managing stress?
- How effective do you think these strategies are?

The HSE has produced a more detailed indicator tool to help identify areas within the workplace that may be adversely affecting staff well-being, and potentially causing stress.

ACTIVITY

Well-being questionnaire

Complete this questionnaire to identify any areas of your work you feel positively about, and those you find more challenging and stressful. You might also like to use this questionnaire with other members of staff.

Instructions: It is recognised that working conditions affect worker well-being, Your responses to the questions below will help you determine your working conditions now, and enable you to monitor future improvements. In order to be able to compare the current situation with past or future situations, it is important that your responses reflect your work in the last six months.

	Never	Seldom	Sometimes	Often	Always
1. I am clear what is expected of me at work	☐1	☐2	☐3	☐4	☐5
2. I can decide when to take a break	☐1	☐2	☐3	☐4	☐5
3. Different groups at work demand things from me that are hard to combine	☐5	☐4	☐3	☐2	☐1
4. I know how to go about getting my job done	☐1	☐2	☐3	☐4	☐5
5. I am subject to personal harrassment in the form of unkind words or behaviour	☐5	☐4	☐3	☐2	☐1
6. I have unachievable deadlines	☐5	☐4	☐3	☐2	☐1
7. If work gets difficult, my coleagues will help me	☐1	☐2	☐3	☐4	☐5
8. I am given supportive feedback on the work I do	☐1	☐2	☐3	☐4	☐5
9. I have to work very intensively	☐5	☐4	☐3	☐2	☐1
10. I have a say in my own work speed	☐1	☐2	☐3	☐4	☐5
11. I am clear what my duties and responsiblities are	☐1	☐2	☐3	☐4	☐5
12. I have to neglect some tasks because I have too much to do	☐5	☐4	☐3	☐2	☐1
13. I am clear about the goals and objectives for my department	☐1	☐2	☐3	☐4	☐5
14. There is friction or anger between colleagues	☐5	☐4	☐3	☐2	☐1
15. I have a choice in deciding how I do my work	☐1	☐2	☐3	☐4	☐5
16. I am unable to take sufficient breaks	☐5	☐4	☐3	☐2	☐1
17. I understand how my work fits into the overall aim of the organisation	☐1	☐2	☐3	☐4	☐5
18. I am pressured to work long hours	☐5	☐4	☐3	☐2	☐1

ACTIVITY continued

19. I have a choice in deciding what I do at work	Never ☐1	Seldom ☐2	Sometimes ☐3	Often ☐4	Always ☐5
20. I have to work very fast	Never ☐5	Seldom ☐4	Sometimes ☐3	Often ☐2	Always ☐1
21. I am subject to bullying at work	Never ☐5	Seldom ☐4	Sometimes ☐3	Often ☐2	Always ☐1
22. I have unrealistic time pressures	Never ☐5	Seldom ☐4	Sometimes ☐3	Often ☐2	Always ☐1
23. I can rely on my line manager to help me out with a work problem	Never ☐1	Seldom ☐2	Sometimes ☐3	Ofter ☐4	Always ☐5
24. I get help and support I need from colleagues	Strongly disagree ☐1	Disagree ☐2	Neutral ☐3	Agree ☐4	Strongly agree ☐5
25. I have some say over the way I work	Strongly disagree ☐1	Disagree ☐2	Neutral ☐3	Agree ☐4	Strongly agree ☐5
26. I have sufficient opportunities to question managers about change at work	Strongly disagree ☐1	Disagree ☐2	Neutral ☐3	Agree ☐4	Strongly agree ☐5
27. I receive the respect at work I deserve from my colleagues	Strongly disagree ☐1	Disagree ☐2	Neutral ☐3	Agree ☐4	Strongly agree ☐5
28. Staff are alway consulted about change at work	Strongly disagree ☐1	Disagree ☐2	Neutral ☐3	Agree ☐4	Strongly agree ☐5
29. I can talk to my line manager about something that has upset or annoyed me about work	Strongly disagree ☐1	Disagree ☐2	Neutral ☐3	Agree ☐4	Strongly agree ☐5
30. My working time can be flexible	Strongly disagree ☐1	Disagree ☐2	Neutral ☐3	Agree ☐4	Strongly agree ☐5
31. My colleagues are willing to listen to my work-related problems	Strongly disagree ☐1	Disagree ☐2	Neutral ☐3	Agree ☐4	Strongly agree ☐5
32. When changes are made at work, I am clear how they will work out in practice	Strongly disagree ☐1	Disagree ☐2	Neutral ☐3	Agree ☐4	Strongly agree ☐5
33. I am supported through emotionally demanding work	Strongly disagree ☐1	Disagree ☐2	Neutral ☐3	Agree ☐4	Strongly agree ☐5
34. Relationships at work are strained	Strongly disagree ☐5	Disagree ☐4	Neutral ☐3	Agree ☐2	Strongly agree ☐1
35. My line manager encourages me at work	Strongly disagree ☐1	Disagree ☐2	Neutral ☐3	Agree ☐4	Strongly agree ☐5

The scores range from 1 to 5. A lower score is likely to indicate a potential problem area.

(HSE et al., 2006)

Managing stress

The HSE suggests the following six steps for managing stress.

1. **Identify the hazard**. The first step is to identify the cause of your stress (the hazard). The well-being questionnaire will help you identify sources of work-place stress, but the HSE stresses that it is also important to review your lifestyle, to see if you can identify any contributing factors. A simple checklist might include:
   - eating on the run, or in a disorganised manner
   - smoking, or drinking excessively
   - rushing, hurrying, being available to everyone
   - doing several jobs at once
   - missing breaks, taking work home with you
   - having no time for exercise and relaxation.
2. **Evaluate the risk associated with the hazard**. Identify what difficulties the stress is causing: for example, an increased sickness record, lower productivity or mistakes at work.
3. **Decide on appropriate control (stress management) strategies**. The HSE notes that 'developing solutions is often the most difficult part of tackling the possible causes of work-related stress. Each workplace and each worker is different, meaning that it is not possible to describe one set of solutions for all circumstances' (HSE, 2008).

   It is important to consider a range of strategies for stress management. These might include:
   - breathing and relaxation techniques
   - ensuring you take a lunch break away from the workplace
   - delegating more effectively
   - discussing your workload with your line manager.
4. **Develop an action plan**. The HSE's next step is to make a plan and stick to it. They identify the benefits of an action plan as:
   - helping you set goals to work towards
   - helping you to prioritise
   - taking your concerns seriously
   - providing something to evaluate and review against.

   The HSE also identifies a number of elements that should be included as part of your risk assessment and action plan:
   - what the problem is
   - how the problem was identified
   - what you are going to do in response
   - how you arrived at this solution
   - some key milestones and dates for them to be reached
   - a date for reviewing against the plan.
5. **Review and evaluate actions**. Review whether your levels of stress have decreased, and whether you are better able to anticipate (and therefore deal with) stressful events or periods. It may take you some time, and testing things out, to find the strategies that work best for you in stress management.

Time management

Time pressures are a common source of stress, and good time management skills are invaluable in reducing stress, as well as increasing your effectiveness as a manager. This section should be read in conjunction with those on other aspects of management practice, including delegation (Chapter 5).

ACTIVITY

How do I spend my time?

List the main tasks you do in your job, grouping together similar activities so you have a maximum of eight tasks.

	Tasks	Fixed or floating?	Level of importance 1 – high, 10 – low
1.			
2.			
3.			
4.			
5.			
6.			
7.			
8.			

- Once you have listed the eight main tasks, identify whether these are fixed or floating.

  Fixed tasks have to be undertaken at set times and usually apply to areas such as:
  - meetings
  - training
  - reviews.

  Floating tasks can be undertaken more flexibly: you could decide at what point during a week you are going to undertake a particular task, such as
  - planning the staff rota for the next month
  - updating budget records
  - ordering supplies.
- Identify how important each of the eight tasks is to your overall role, ranking the level of importance 1 (high) to 10 (low).

REFLECT

Look at your answers to the activity above and consider these questions.

- Which of the eight tasks takes up the most time?
- Do the most important tasks take up the majority of your time?
- Are there any tasks, or elements of tasks, that you could delegate? If not, can you justify why not?
- Do floating tasks regularly remain undone?

Can you now identify any areas of time management that you could improve and manage more effectively? Identify up to three things you could do to improve your time management.

Time management tips

- Prioritise tasks by developing a system that works for you, such as this.

Group A	Do today
Group B	Do this week
Group C	Deadline beyond this week – schedule time to complete task in diary

Go through your in-tray and list of jobs, and place in one of the three groups. You may have separate trays for each group. The next time you handle an item, deal with it. If you continue to re-schedule the items in Group C, consider whether they are really essential; if so, move them into Group B, and complete within the week.

✔ Be assertive in protecting your time, saying no and delegating.

✔ Deal with emails only once or twice a day: for example, at midday and 4 p.m.

✔ Alternatively, have an 'email-free' day. Let people know that you will be doing this, and when they might expect a response from you (add to the auto-response on your email).

✔ Organise your time to fit your own style and working patterns. For example, if you are not a 'morning person', it may be more productive to schedule an appraisal meeting for late morning or afternoon; if you are a 'deadline person', acknowledge this and ensure you schedule time before a deadline to complete tasks. Be aware, however, that if an emergency arises, you may miss the deadline!

✔ Make deadlines for tasks you do not enjoy and schedule time in your diary to complete them.

✔ Schedule a regular, quiet time when you can deal with tasks such as reading new policy guidance or filing. This may simply mean shutting your office door and diverting your phone for a few hours.

✔ Always consider:

- Do I need to deal with this?
- Do I need to deal with this now?

A final note

Remember to have fun with your CPD too!

NVQ/SVQ

This chapter will help you to provide evidence for the following units:

- **Mandatory Units A1**

Work Products you may generate and either include in your portfolio or show your assessor to demonstrate your skills and knowledge are:

- **records of appraisals and development plans**
- **learning logs from training events/ conferences attended**
- **records of reflection from other learning activities**
- **records of re-registration with professional bodies, such as GSCC.**

Figure 10.10 CPD can really help your career take off!

References

CIPD (2008b) *What is CPD?* London: CIPD
GSCC (2002) *Code of Practice for Social Care Workers and Code of Practice for Employers of Social Care Workers*, London: GSCC
GSCC (2007) *Guidance notes on how to renew your registration*, London: GSCC
GSCC (no date) *Post-Registration Training and Learning (PRTL) requirements*, London: GSCC
HPC (2004) *Continuing Professional Development – Consultation Paper*, London: HPC
HPC (2006) *Continued Professional Development and Your Registration*, London: HPC
International Stress Management Association, HSE and ACAS (2004) *Working together to reduce stress at work: A guide for employees*, International Stress Management Association
HSE (2008) *Management Standards for Stress: Step 3* www.hse.gov.uk/stress/standards/step3/index.htm (accessed 03.08.08)
HSE, Department for Work and Pensions, Cabinet Office and The Work Foundation (2006) *Ministerial Taskforce on Health, Safety & Productivity: The Well Managed Organisation Diagnostic Tools For Handling Sickness Absence*, www.hse.gov.uk/services/pdfs/diagnostictools.pdf (accessed 03.04.03)
Rothwell, A. and Arnold, J. (2005) 'How HR professionals rate continuing professional development', *Human Resource Management Journal*, Vol 15, No. 3:18–32. Cited in Megginson, D. and Whitaker, V. (2007) *Continuing Professional Development* (2nd edition), London: The Chartered Institute of Personnel and Development
Skills for Care (2006) *Leadership & Management Strategy: A Strategy for the Social Care Workforce*, Leeds: Skills for Care

Useful reading

Alsop, A. (2000) *Continuing Professional Development: A Guide for Therapists*, Oxford: Blackwell Science Ltd
Megginson, D. and Whitaker, V. (2007) *Continuing Professional Development*, (2nd edition), London: Chartered Institute of Personnel and Development
Moon, J. (1999) *Learning Journals: A Handbook for Academics, Students and Professional Development*, (2nd edition), Abingdon: Routledge
Pearce, R. (2006) *Profiles and Portfolios of Evidence: Foundations in Nursing and Health Care*, Cheltenham: Nelson Thornes

APPENDIX 1

Case study answers

Chapter 2: Leadership and management

Case study 1 page 18

CASE STUDY

Leadership styles

Consider the following scenarios and identify which leadership style is being used.

At a team meeting, ***Angela****, the manager, allocates half an hour for the group to discuss ideas for the home's summer fete. Individuals come up with a range of ideas. Angela thanks everyone for their contributions and says she will confirm the plans for the fete at the next team meeting.*

A few team members are discussing the shift rota and trying to arrange some swaps. ***Bernie****, the manager overhears the conversation and says all requests must come through him for his decision, and he reminds the team members that his decision will be final.*

The residential home you work in has recently been inspected and a few areas for improvement have been identified. ***Carmen****, the manager, has placed a suggestion box in the reception area so that staff, service users and family members can contribute their ideas of how to improve the areas identified. Carmen has also had three 'surgeries' in evenings and weekends, to give people the opportunity to speak to her directly about their ideas. Next week, you will be involved in the staff and residents' meeting where the ideas will be discussed and the best ones selected.*

It is the end of October and, at a team meeting, several people ask about leave arrangements over the Christmas and New Year period. ***Dougie****, the manager, tells everyone to sort it out among themselves and let him know the shift rota once they have agreed.*

The leadership styles being used here are:

- **Angela** is using a democratic style, consulting the team and recognising their ideas and viewpoints, but making the ultimate decision herself.
- **Bernie** is using an authoritarian style, controlling all decisions without consultation and holding the final decision.
- **Carmen** is using a participative style consulting her team and involving them in the decision making and then integrating these ideas into her final decisions.
- **Dougie** is using a laissez-faire style devolving the decision making and allowing groups or individuals to decide for themselves.

Case study 2 page 20

CASE STUDY

Using emotional intelligence

Elaine is a care assistant. She has recently gone through a difficult divorce and is the sole carer for her two children, aged 8 and 11. The school holidays are coming up and there is nobody to assist with her childcare. Elaine has come to her manager, Tom, to discuss this. She is rather aggressive and demanding.

Tom is sympathetic to her problem. He recognises that Elaine's aggression is a result of her anxiety about the school holidays. He quickly reassures Elaine and tells her he is sure something can be arranged. Tom agrees to look at the options and talk to her again by the end of the following day.

The next day, Tom meets with Elaine and makes a number of suggestions. He suggests that she could take five weeks unpaid leave during the summer holidays, but spread this across the whole year so that she will still receive a salary each month. He also offers the option of reducing the hours of her working day during term time, if it would help Elaine to take her children to and from school more easily. He tells Elaine to think about these options and come back to him when she has made a decision.

- *In what respects is Tom an emotionally intelligent leader?*
- *What might the outcome of his suggestions be for the organisation and Elaine?*
- *What would you have done?*

(Adapted from Mason, 2003)

These questions can be answered as follows:

In what respects is Tom an emotionally intelligent leader?
Tom has empathised with Elaine's circumstances and communicated clearly with her, reassuring her that he will try to assist. He is aware that Elaine's aggression is a result of her anxiety, rather than personally directed at him.

What might the outcome of his suggestions be for the organisation and Elaine?
Tom has offered Elaine some possible solutions to her difficulties. By being flexible and offering options, Elaine is more likely to be able to continue to work at the residential unit. She is also likely to be happier and more motivated if her child care difficulties are resolved.

Tom is more likely to retain a valuable member of staff. It will prevent the need to recruit and train another member of staff, and be less disruptive to residents.

Chapter 3: Working with service users, families and carers

Case study 1 page 33

CASE STUDY

Read through the four case studies below and identify which of Erikson's stages of development and conflicts each individual is going through.

***Billy** is a young man with learning disabilities. He is very sociable and has a good group of friends, and has had several girlfriends in the past, but only for short periods. He tells you that he is tired of dating different girls and wants to have one special 'forever' girlfriend.*

***Florence** can't wait for her grandchildren's next visit. She loves to spend time telling them stories about their mother when she was their age. She talks about the home they lived in and holidays they enjoyed together as a family. Sometimes, she says she could have been a famous singer, but she acknowledges that, if she had kept trying to pursue that dream, she 'wouldn't have had time to have your mum!'*

***Stanley** is an 82-year-old widower who lives in a sheltered housing scheme. Stanley is in relatively good health, but is frail. He rarely goes out and does not take part in activities offered, staying in his flat instead. Other residents in the scheme say 'it is better that way' because all he does is moan about how difficult his life has been, and how he has 'had things worse' than most people.*

***Mia** is 33 and went through a residential drug rehabilitation programme for the second time three months ago. She is currently attending the after-care day service program offered by the unit. She is making good progress in getting things back on track. She has just booked an appointment with a careers adviser to help her to decide what to do 'with the rest of her life'.*

The stages of development being shown here are:

- **Billy** is at the Young Adult stage with the conflict of intimacy vs. isolation
- **Florence** is at the Middle-age Adult stage with the conflict of generativity vs. stagnation
- **Stanley** is at the Older Adult stage with the conflict of integrity vs. despair
- **Mia** is at the Adolescent stage with the conflict of identity vs. role confusion

It is important to remember that when assessing an individual's stage of development that they may not always relate to an individual's age. While Billy, Florence and Stanley appear to be following Erikson's stages of development/conflicts more associated with their age, Mia is at a stage more associated with a teenager. Experiences that individuals go through have a direct affect on their age/stage of development. Never make assumptions based on chronological age. Always look at the individual needs and experiences of the people you are working with to ensure their needs are met.

Case study 2 page 35

CASE STUDY

Frank

Frank is a widower who moved into your care setting last year. His wife had died six months earlier and Frank could no longer cope at home. He has early stages of dementia. Puddles, the cat which belonged to Frank's wife, also moved into the home with him. Puddles provides great comfort to Frank and has become a favourite with the other residents. You arrive at work one morning to discover that Puddles has been run over and has died.

- *How will you support Frank through this loss?*
- *How will you support other residents?*

Simone

Simone has learning difficulties and lives in a semi-independent unit which is part of the service you manage. Simone's younger sister Anya visits her frequently and they are very close. Anya is sitting her 'A' levels in a few months and, assuming she is successful in her grades, will be going away to university in September. This means Simone will see Anya much less often.

- *How will you work with Simone and the staff team to support her in the periods before and after Anya goes to university?*
- *How might you support Frank and Simone at each of the stages of grief identified by Kiüler-Ross?*
- *What strategies would you use at each stage and why?*
- *What other factors might you need to keep in mind in each situation?*

Frank

How will you support Frank through his loss?

- Acknowledge that the loss will be significant to Frank – the cat belonged to his wife and offered him comfort and companionship.
- Comfort and reassure him.
- He may have difficulty coming to terms with the loss, which means you may need to repeat and reinforce to him that his cat has died.
- Give Frank time to talk about how he is feeling – this may also bring back thoughts and feelings he had when his wife died.
- Do not immediately suggest replacing the cat.
- Listen to him.
- Be calm, patient and sympathetic.

How will you support the other residents?

- Acknowledge they will feel the loss also.
- Be aware that it may make them think of their own mortality.

Simone

How will you work with Simone and the staff team to support her in the periods before and after Anya goes to university?

- Prior to Anya going to university spend time talking to Simone about where her sister is going, why she is going and when she will be returning.
- Be positive and constructive when talking to Simone about her sister going to university – this could be related to a time when Simone was involved in new activity and how she had benefited from it.
- Acknowledge this will be a difficult time for Simone; she may have difficulty understanding why her sister is going/gone away. Staff may need to repeat and reinforce this on many occasions – they will need to remain calm and patient while Simone adjusts.
- Possibly arrange a visit to the university so she can see where her sister will be, or once she is at university arrange a visit to see her there.
- Encourage Simone to stay in contact with her sister via phone, letter or e-mail – this is an area that staff can help her with if she encounters difficulty.
- Staff will need to understand that Simone will go through the grieving process when her sister goes to university and will need to be given support.

Frank and Simone

How might you support both Frank and Simone at each of the stages of grief identified by Kübler-Ross? What strategies would you use at each stage and why?

- Acknowledge that they will go through each stage of grief at their pace.
- Do not try to rush them through a stage before they are ready – some people will move quite quickly through stages others will take longer depending on their coping mechanisms and experiences. Don't tell them to 'pull themselves together'.
- You may need to gently remind them of their loss – pretending it has not happened will not change the situation.
- Do not take anger or frustration personally – it is the situation they are going through that is making them respond in a particular way.
- Be patient and sympathetic.
- Do not try to replace their loss – i.e. replacing Frank's cat. It will not be the same and could add to the upset and distress.
- Repeat and reinforce that they are not to blame – i.e. Simone has not done anything wrong to make her sister go away.
- Acknowledge that they are feeling sad and lonely.
- Once they reach the Acceptance stage you could make suggestions about something new they could become involved in.

What other factors might you need to keep in mind in each situation?

- You will need to be aware of the level of understanding of each individual.
- You will need to be aware of past experiences.
- The loss for Frank is also associated with his wife and this will be a significant factor to take into account.

Case study 3 page 40

CASE STUDY

Approaches to risk management

'My friend Chris works all day and can only come and see me in the evening. He asked if this was OK, and the staff assured him that it was fine – there are no "visiting times" here – after all it's my home. They don't just let people traipse in and out though as they have to look after our safety too.'

(CSCI, 2005, Care Homes for Older People: National Minimum Standards)

- *Do you think the practice above demonstrates a safety-first or risk-taking approach to visits from relatives and friends?*
- *If you applied the safety-first approach to relatives and friends visiting service users in your setting, what type of procedures and rules would you have in place?*

If you applied a risk-taking approach, what types of procedures and rules would you have in place?

Do you think the practice above demonstrates a safety-first or risk-taking approach to visits from relatives and friends?

The practice by staff demonstrates a risk-taking approach to visits from friends and relatives. It acknowledges the service users rights to make choices in relation to their lives and social networks.

If you applied the safety-first approach to relatives and friends visiting service users in your setting, what type of procedures and rules would you have in place?

Safety-first rules for visitors may be:

- Vetting visits from friends and relatives.
- Only having visits that are pre-planned.
- Staff making contact with service user friends and relatives.
- Staff making all of the arrangements and setting the times for visits.
- Staff supervising visits.

If you applied a risk-taking approach, what types of procedures and rules would you have in place?

Risk-taking approach rules for visitors may be:

- Staff getting to know service user friends and family, which enables more freedom of access to the setting.
- Empowering and encouraging service users to arrange visits with their friends and family.
- Give choices when and where service users have contact with friends and family.
- Allowing unplanned visits from known friends and family members, although before and after times may be set, i.e. no visitors before 9 a.m. and after 10 p.m. Staffing and needs of other service users still need to be considered.
- Staff not supervising visits with friends and family.

Case study 4 page 40

CASE STUDY

Reviewing risk

Phillip has been a resident in your setting for four years. He has a learning disability and communication difficulties, which can make it quite difficult for him to express his views to others. Phillip has been very independent; he has been working in a local supermarket two days a week, taking himself to and from work. He also enjoys going to rugby matches with a member of his family. His key worker Justin has a good working relationship and communicates well with him. Recently Phillip has started to have 'fits', which are being investigated by the local hospital. This has affected the level of independence he has, impacting on his ability to work and go on social outings, which Phillip is finding very difficult to understand.

You need to review the initial risk assessment regarding Phillip's circumstances and have a review meeting.

Answer these questions.

- *How will you support Phillip during this period?*
- *How will you help to understand the changes to his health, work and social life?*
- *How will you involve Phillip in the risk assessment and review meeting processes?*
- *What do you need to do prior to the review meeting?*
- *Who else will you involve in the review? Why?*
- *What strategies and supports could be put in place to ensure Phillip retains as much independence as possible?*
- *Have you addressed each of Titterton's steps?*

How will you support Phillip during this period?
You will need to understand that Phillip will be feeling frustrated and angry at the changes he is encountering and be understanding and patient with him.

How will you help to understand the changes to his health, work and social life?
Explain why the changes to his health have affected his work and social life. Keep repeating and reinforcing this.

How will you involve Phillip in the risk assessment and review meeting processes?
Go through the process of risk assessment with Phillip so he can see it as a process to enable him to do all the things he enjoys rather than stop him. Recognise for his safety there may be some restrictions but look at ways to address these, involving Phillip at all times in the decision-making process.

What will you need to do prior to the review meeting?
Write a detailed risk assessment before the review outlining the risk and potential ways to deal with them.

Who else will you involve in the review? Why?
Involve Phillip's key worker when writing the review as he knows Phillip well and will be able to contribute constructive suggestions. Also include Phillip, his family members, key worker and someone from the funding authority to the review.

What strategies and supports could be put in place to ensure Phillip retains as much independence as possible?
Make sure that Phillip is listened to and his choices are considered as much as possible. Look at ways that he can still attend the rugby matches – this may involve a staff member attending along with a family member. Possibly arrange transport to and from work.

Case study 5 page 42

CASE STUDY

Family pressure

A family member pressurises members of your staff team to share personal health information about a service user that the service user doesn't want to be shared with them.

- *How would you manage this situation with staff, the service user, and family member, ensuring that productive relationships are maintained?*

Shift team problems

It has come to your attention that there are some problems in staff not sharing relevant information and updating other team members fully at shift handovers. This has caused some gaps in continuity of care between shift teams.

- *How would you manage this situation with the shift teams?*
- *What strategies could you adopt to minimise the risk of this situation occurring again?*

Family pressure

How would you manage this situation with you staff team to share personal health information about a service user that the service user doesn't want to be shared with them?

Explain to the family member that staff are not allowed to share information as they have to respect service user confidentiality. Highlight that staff are not being obstructive or difficult but are working in line with the setting's confidentiality policy and are meeting legislation requirements. Reinforce the message that by all working together and ensuring the service user rights are being promoted is beneficial for everyone.

Shift team problems

How would you manage this situation with the shift teams? What strategies could you adopt to minimise the risk of this situation occurring again?

- Have a communication book that staff complete with important messages.
- Staff to sign and date the book as they read the messages.
- Have a handover sheet that the shift leader is responsible for filling in and passing on to the next shift leader. Include headings such as service users needs, incidents, concerns, visits etc.
- Including a 15-minute cross over time on shifts so that staff from both shifts are in the setting at the same time.
- Do periodic checks on the message book and handover sheets to make sure they are being completed clearly and accurately.
- Raise the issue of good communication during staff meetings and supervision sessions.

Chapter 4: Workforce planning and development

Case study 1 page 54

CASE STUDY

Interviewing

You are taking part in an interview panel with another manager from your organisation, recruiting an assistant carer for the residential home she manages. Your organisation has a set of interview questions for this job role that have been used for the last two years.

During one interview, you become aware that the applicant seems to anticipate the questions you are about to ask, and answers each question covering the elements you have in the organisation notes for the 'ideal response'. You are aware that the applicant has a friend who already works for the organisation.

- *Why would this situation give you cause for concern?*
- *What action might you take? Who else in the organisation could you seek guidance from?*
- *How could this type of situation be prevented in the future?*

Why would this situation give you cause for concern?
This would give cause for concern as candidate may have been coached to give 'exact' answers to the set questions. This will make it difficult to judge suitability for the job role as you will not be able to differentiate between candidates knowledge and experience and what they have been instructed to say.

What action might you take? Who else in the organisation could you seek guidance from?
Ask the candidate to give a specific example of an experience they have had in their professional career that relates to one of the questions you have asked. You could talk to someone in personnel or a senior manager for guidance.

How could this type of situation be prevented in the future?
Don't use the same set of questions for every interview and ask candidates to reflect on an area of their practice.

Case study 2 page 57

CASE STUDY

Offereing new challenges

Amina has worked in your setting for five years; she has completed her NVQ2 and NVQ3 during this time. She is key worker for two service users. She knows and understands the service users' needs and works well with them. She has been a caring, committed member of your team who has taken on additional responsibilities and is able to use her own initiative.

In her recent supervision, she told you that she was not enjoying her job as much as there are no new challenges and as a result she is feeling demotivated.

You have a new member of staff starting in your setting. The staff member is also new to residential work.

- *How could you include Amina in the induction process of the new member of staff?*
- *How might this help the new member of staff?*
- *How might this benefit Amina?*
- *How might Amina's involvement in the induction benefit the setting?*

How could you include Amina in the induction process of the new member of staff?
Amina could be allocated as the new staff members' mentor. This would make her involved in planning and delivering aspects of the induction. She would also have the opportunity to take the staff member through aspects of the induction standards.

How might this help the new member of staff?
The new staff member would have an experienced member of staff to offer guidance and support, as well as an additional person (in addition to line manager) to talk through any issues. The new staff member may feel more able to raise concerns or areas they feel unsure about with someone that does not have line management responsibility for them.

How might this benefit Amina?
This position would acknowledge Amina skills and experiences and allow her to share her knowledge and experience. She will receive a new challenge and be able to develop supervisory and planning skills. It will also give her the opportunity to look at different career options, for example management, assessing etc.

How might Amina's involvement in the induction benefit the setting?
Amina will be happier, which will make her feel more motivated and valued which makes it more likely that Amina will stay in setting. It also shares the work load, meaning the manger does not have to do everything. New members of staff also benefit from an experienced staff members knowledge and expertise, while other staff will be encouraged to develop their skills.

Case study 3 page 58

CASE STUDY

Planning inductions

You have recently appointed two new deputy managers within your care home. Jessie has been working in your setting for five years and has been acting deputy for the last six months. Robert is new to your setting, organisation and management. He has worked in residential settings for ten years. He was a key worker and shift leader in the setting he worked in previously. You are responsible for planning their induction and taking them through the management induction standards.

Answer the following questions:

- *Will you plan the same induction for Jessie and Robert?*
- *If yes explain why, if no why not.*
- *How will you incorporate the induction standards?*
- *What are the benefits of having induction standards? Are there any disadvantages?*
- *How can Jessie and Robert support each other during this process?*

Will you plan the same induction for Jessie and Robert? If yes explain why, if no why not.
Jessie and Robert will need different induction plans that are designed to meet their individual needs. Jessie has worked with the organisation for several years and will know the structure and policies and procedures. However Robert is new to the setting and organisation and will need to be given time to get to know and understand these.

How will you incorporate the induction standards?
Try to view the Standards creatively and avoid taking a purely 'tick box' approach to meeting them. Consider using a range of methods to achieving them, and consider which approaches will be most appropriate for Jessie and Robert.

What are the benefits of having induction standard? Are there any disadvantages?
The Induction Standards will be a good starting point for both staff members. The standards will ensure that you are able to plan a detailed induction package for both Jessie and Robert. It is important not to assume knowledge with any staff member, for example Jessie has worked in the organisation for five years but she is new to management. It can also be difficult for staff when they are promoted within a team that they have worked in. Jessie will now have line management responsibility for staff that have until recently been peer colleagues. The standards can help you develop Jessie and Robert's management skills and give them the opportunity to seek clarification on areas that they are unclear about.

How can Jessie and Robert support each other during this process?
Jessie and Robert will also be able to support each other. They are both new to management, although both have a lot of experience in the health and care sector. Jessie will know and understand the organisation and will be able to share this information with Robert, while Robert, being new to the organisation, may be able to take a more objective approach when looking at systems, structures, procedures and staffing issues.

Case study 4 page 62

CASE STUDY

Planning for supervision

Consider the following members of staff and answer the questions below for each of them.

- o *Staff Member 1 is a new member of staff who is not very confident and lacks initiative.*
- o *Staff Member 2 uses supervision to gossip about others in the team.*
- o *Staff Member 3 uses supervision only to discuss their personal issues.*
- o *Staff Member 4 does not value supervision and sees it as a waste of their time.*

- *How are you going to work with each of the following staff members to make supervision productive?*
- *How will you set agendas and keep the supervision focused?*
- *Will you have common areas on each agenda? Why/why not?*

How are you going to work with each of the following staff members to make supervision productive?

- Have an agreement in place that the staff member is involved in drawing up.
- Treat each staff member as an individual – at times a manager has to be flexible and adaptable in their approach to ensure they can work effectively with all staff.
- Be clear on why you are having the supervision and the benefits to the staff member.
- Set time limits.

How will you set agendas and keep the supervision focused?
Have an agreed, set agenda to help keep staff members focused on work-related issues – discuss with staff member areas that they feel should be included in 'their' supervision. If you disagree with items then give a clear explanation why.
Have an allocated slot where they can discuss 'other issues' – be clear on timing for this so that it doesn't dominate. Also include any other business at the end of the meeting as this gives staff time to raise areas/issues that they don't think fit in with set items.

Will you have common areas on each agenda? Why/why not?
Having common agenda items in a setting can benefit you as a manager but you need to make sure they are appropriate to all staff. Staff will not feel valued if half of the 'fixed' agenda items do not relate to their job role or needs. You may have different headings for different staff roles.

Case study 5 page 62

CASE STUDY

Dealing with undermining behaviour

You have recently started supervising a member of staff who has worked in your organisation for the same length of time as you. You both started at the same grade, but you have chosen to pursue a career in management, while she has remained at the same grade. She does not feel that she needs to be supervised and has made it clear that you 'can't teach her anything'. You have held two supervision sessions with her, where she has attempted to undermine your role, and has criticised any comments and suggestions you make.

- *Why do you think she may be acting in this manner?*
- *What strategies can you use to make supervision more constructive?*
- *Could her behaviour affect the team? How?*
- *Would you seek advice or support from anyone else regarding the situation?*

Why do you think she may be acting in this manner?
The staff member may be feeling threatened and undermined that they have not developed at the same pace as you. She might be reflecting on her life achievements and not feeling that she has 'succeeded' and may feel that other staff members see her as a failure.

What strategies can you use to make supervision more constructive?
Draw up an agreement and agenda for supervision and show that you value the staff member's experience and skills. Find out if there are areas of their practice or knowledge they would like to develop – if this can be facilitated then put it into action.

Could her behaviour affect the team? How?
As an experienced member of staff she could have a significant affect on the staff team. She can undermine your role and cause unrest in the team, which is disruptive and destructive.

Would you seek advice or support from anyone else regarding the situation?
You can talk through the issues with your own manager/head of service and also to peer managers for support and sharing strategies. Do not share your frustration with other staff members; remember to remain calm and professional when dealing with the difficult member of staff.

Case study 6 page 65

CASE STUDY

Giving feedback

You have carried out an observation of a member of staff. The member of staff was supporting a service user for whom they are keyworker, to complete the 'My Plan' booklet for the service user's review in two weeks' time. The service user has learning difficulties and has some difficulties communicating verbally.

The staff member sat in the sitting room with the service user and had a friendly, warm approach. She sat alongside the service user so that they could see the booklet. She explained what they were doing and why they were doing it. You observed that she didn't check the service user had understood this. She then went through the booklet and asked questions. As the service user attempted to answer a number of questions, the staff member interrupted and answered for him.

- *Which areas of practice would you identify as good, and which areas need ongoing development?*
- *How would you give feedback to the member of staff? What would you say? When and where would you say it?*
- *How will you support the member of staff to develop in the areas identified?*

Which areas of practice would you identify as good, and which areas need ongoing development?
Positive areas of practice include:

- The staff member has a warm, friendly approach with the service user.
- She sat with the service user so they could see the book and explained to the them, what the book was and why they were going to fill it in.

Areas to be developed include:

- Do not assume communication has been understood – ask questions to check.
- Listen to service users and give them time to answer.

How would you give feedback to the member of staff? What would you say? When and where would you say it?
Give feedback during supervision, not in front of other staff or service users. Be sure to point out the good practice. Ask the staff member to identify if there is any area of their practice that they feel they could develop. You can also ask them to give an example of a time that they have not been listened to or talked over and ask them to reflect how they felt.

Link to their good practice to show how by developing areas of their work it will complement what they do and enhance their interaction with service users.

How will you support the member of staff to develop in the areas identified?
Use supervision to help the staff member reflect on their practice, looking at training courses related to communicating with service users. If they are doing their NVQ link it to relevant units. Be sure to give feedback when the staff member communicates well with service users and show how this benefited them.

Chapter 5: Common challenges in management

Case study 1 page 78

CASE STUDY

Team-building scenarios

Consider these two scenarios and answer the questions below.

Scenario one *You work in a residential care home for people with learning difficulties. The home has recently been inspected and criticised for failing to take a person-centred approach to care planning. All teams within the organization must produce an action plan to improve their performance in this area.*

Scenario two *You work in a residential care home that provides care and support to adults recovering from drug and/or alcohol addiction. Until six months ago, you had a very stable and experienced team with close working relationships. The team had worked together for over two years until two members left the organisation within a few weeks of each other. Two new members of staff were recruited to the team fairly quickly, but previous team members obviously miss their former colleagues and continue to maintain their friendships outside of work. The new team members seem to be struggling to integrate fully into the team.*

- *How might team members be feeling in each case?*
- *How might Tuckman's and Belbin's theories inform your understanding of the situations?*
- *Choose one of the scenarios and plan a team event and activities to meet the needs of the team. Present this to a group of colleagues, justifying your plan and choice of activities.*

Scenario 1

How might team members be feeling in each case?
The team are likely to be feeling disappointed, and possibly demotivated or angry. Team members may also recognise that the comments are fair and want things to improve.

How might Tuckman's and Belbin's theories inform your understanding of the situations?
The stage of development which the team is at is likely to affect how the team cope with the situation. If the team are at 'norming' or 'performing' stages they may find it easier to work together to resolve the issues, and to support each other in the process. The team may also move to 'storming' stages if conflict arises in deciding how to move forward.

You may consider Belbin's role in identifying roles which individuals can play in implementing changes and moving teams forward. For example, roles associated with thinking and problem solving are likely to be most helpful in finding solutions to the areas identified for improvement. Team workers will be important in helping the teams to move forward.

Scenario 2

How might team members be feeling in each case?
Team members may be feeling unsettled, and the new team members are also likely to be feeling excluded and isolated.

How might Tuckman's and Belbin's theories inform your understanding of the situations?
The team appears to have been at the 'performing' stage and may now be moving into a 'storming' phase as relationships are reviewed and the new team members become integrated into the team.

You may be relying on the team members who meet the roles identified by Belbin as associated with people and feeling, to try to develop a sense of cohesion within the team and to be optimistic about the potential of the new team.

Case study 2 page 82

CASE STUDY

Change Diagnostic Tool

Read these statements, which have been made by members of staff in reaction to being told about a change that needs to take place.

- *'Will I be made redundant?'*
- *'Here we go again!'*
- *'I think this is going to work well'.*
- *'How much extra work is this going to be?'*
- *'It's too complicated – how are we supposed to do this?'*

Identify which stage of the change process you believe the member of staff is at and how you might respond and work most effectively with the member of staff.

Will I be made redundant?
Shock and denial

Here we go again!
Shock, denial and resistance

I think this is going to work well.
Acceptance or commitment

How much extra work is this going to be?
Resistance and acceptance

It's too complicated – how are we supposed to do this?
Resistance or exploration

Case study 3 page 86

CASE STUDY

Managing conflict

For the past year, you have been managing a cohesive team of individuals who work well together. A new member of staff joins your team, and within a short period of time you notice that team dynamics have changed and there appears to be tension between individuals that was not present previously.

One team member informs you that the new member of staff spends a lot of time gossiping to other staff, particularly criticising your management practice. Another team member informs you that they are feeling very unsettled and are considering leaving because of the disruption and tension caused by the new team member.

- *What do you think might be happening here and why?*
- *What immediate action might you take?*
- *Would you approach the new team member individually? If so, how and why?*
- *How will you work with the team as a whole to resolve the situation and restore good team working?*

What do you think might be happening here and why?
The arrival of a new team member appears to have had an unsettling effect. This may be a natural transition period but the comments from the other team members are concerning and suggest that there are other factors which need to be considered. The new member of the team may be having difficulty settling in, or may be struggling with the work. It may be that the new worker is finding the work more difficult than expected, or that the job as a whole does not meet their expectations.

What immediate action might you take? Would you approach the new team member individually? If so, how and why?
You should consider speaking with the new team member as soon as possible. This should not be done in front of other team members, but privately. You may wish to begin by asking how the team member is settling in and if they are finding anything difficult or not expected. You should then raise your concerns if the team member does not raise anything themself.

How will you work with the team as a whole to resolve the situation and restore good team working?
You will need to support the team in working with the new team member, and might suggest strategies for dealing with any distractions and gossip. You may identify a member of the team to act as a 'buddy' or 'mentor' to help the new team member to settle in, and to understand the accepted practices and behaviour within the team.

Case study 4 page 87

CASE STUDY

Motivation and satisfaction in social care

Look at the three figures below from a National Survey of Care Workers completed for Skills For Care (TNS, 2007).

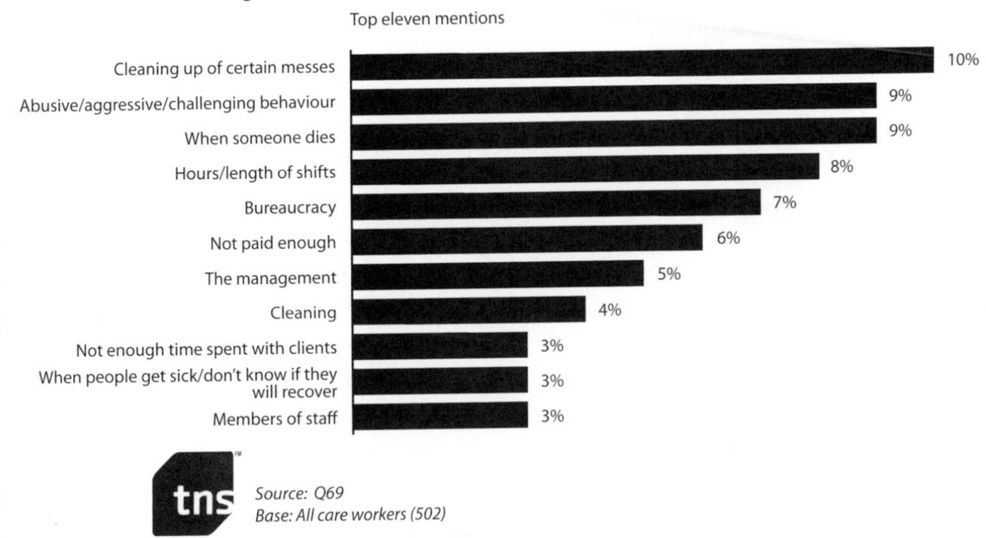

Figure 5.12 The worst things about work

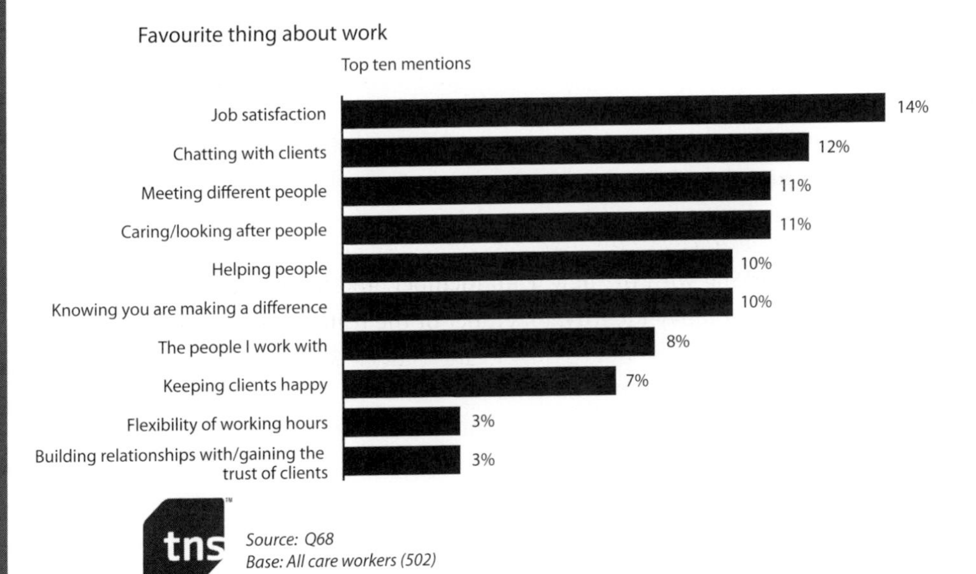

Figure 5.13 Favourite things about work

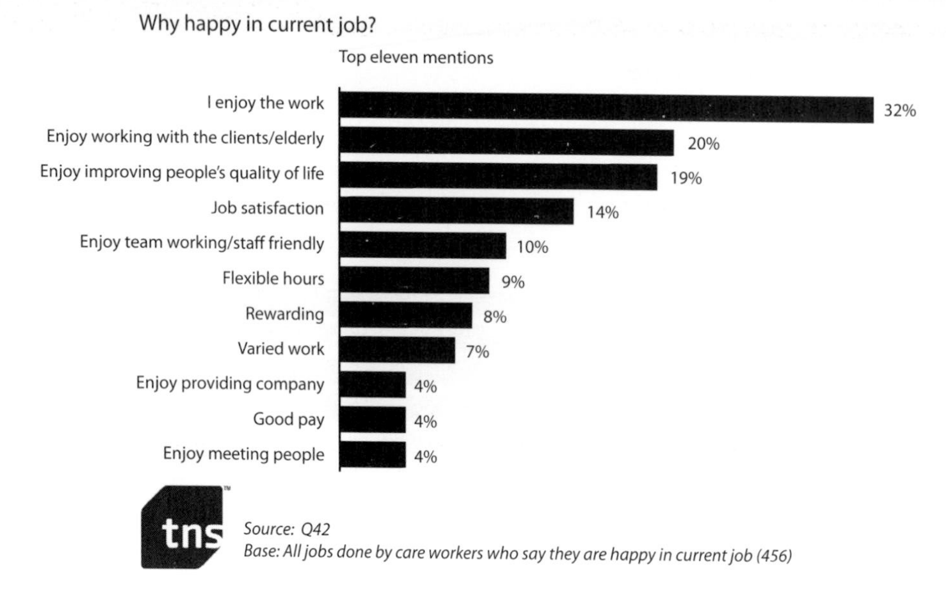

Figure 5.14 Reasons to be happy in current job

- *Do the responses from the survey support McClellands or Hertzberg's theories of motivation?*
- *Which theory, if any, do they support most strongly?*
- *How could the theories and results from the survey inform your practice in motivating staff?*

Do the responses from the survey support McClellands or Hertzberg's theories of motivation?
In terms of McClelland's theory, most care workers appear to fit in the achievement-motivated category (likes doing a good job) and affiliation-motivated (team working).

Which theory, if any, do they support most strongly?
However, the survey seems to support Hertzberg's theory most closely, as the majority of care workers indicate that they do this type of work, and achieve most satisfaction from working with people and the nature of the work (helping to improve the quality of life of people).

How could the theories and results from the survey inform your practice in motivating staff?
The theories and results from the survey emphasise the importance of team-working and of staff feeling they are helping to improve the quality of life of people. Staff are more likely to be de-motivated where there is conflict within teams. High turnover of staff or vacancies may restrict the time staff are able to spend with service users and lead them to feel frustrated and de-motivated about the quality of work they are able to do.

It is important for staff to understand how procedures and any changes to the service will contribute to the improvement of quality of life for service users.

Although staff will obviously want to be paid a reasonable salary for their work, and to have good conditions of service (annual leave etc.), they are also likely to be motivated by receiving positive feedback and thanks for their work.

Also consider how staff may be given recognition where they are involved in exemplary work. There may be regional schemes such as 'Care Worker of the Year' awards; Skills for Care run a 'Care Ambassadors' programme, or your organisation may have, or wish to develop, a scheme to recognise workers achievements.

Case study 5 page 94

CASE STUDY

Action planning

Consider these scenarios.

Scenario one *You have heard some of Paula's colleagues refer to her as 'grumpy' on morning shifts on several occasions. Today a resident mentioned to you that Paula had 'really upset' another resident, by speaking to her harshly over breakfast.*

Scenario two *Rosemarie is excellent at interacting with service users and their family members, but is not very organised in her paperwork. She has recently 'mislaid' two important documents, one relating to a service user and the other a receipt for a piece of equipment.*

Scenario three *You have noticed that Brendan has begun to look more untidy, and also tired at work. Recently he has worn clothes that are torn and stained. Today Brendan fell asleep while sitting and talking with a service user.*

For each of the scenarios, answer these questions.

- *What are possible reasons for the behaviour?*
- *Are you concerned about the behaviour and, if so, why?*
- *What changes would you expect to see? And by when?*
- *How could you support the worker to make the changes required?*

What are possible reasons for the behaviour?
They may be having personal difficulties which are impacting on their work, struggling with their workload or not gaining satisfaction from their work.

Are you concerned about the behaviour and, if so, why?
Each scenario is concerning. In each scenario you would need to speak with the individual concerned and try to establish causes for their behaviour/appearance.

What changes would you expect to see? And by when?
The course of action you decide to take will depend to a great extent on the cause of the difficulties, however you should be clear about your expectations and that of the organisation. You should also identify any action required and agreed, and set a date to review the situation. You should also make a clear record of the discussion and agreed plan, which should be signed by the member of staff and you, with a copy given to the member of staff.

How could you support the worker to make the changes required?
Supervision is likely to be one of the primary methods of support for the workers in each scenario. You may also identify training and development activities for the workers within the action plan you complete with the member of staff.

You should also consider whether you need to provide advice or refer the workers to other departments or services. For example you may refer them to human resources or suggest they see their GP if there are physical or emotional health issues.

You should also ensure that you provide positive feedback where you see change taking place, and review the workers action plan and progress on a regular basis.

Chapter 6: Delivering quality services

Case study 1 page 102

CASE STUDY

Staying user-focused

Brierdene residential home for older people has recently completed a survey of a number of different aspects of the care offered within the home. The area which came out worst in terms of service user satisfaction was mealtimes. Several service users have described it as 'regimented' or 'like a conveyor belt'.

Identify changes you might introduce to mealtimes to overcome this, using the characteristics of quality care as defined by service users in the research by Innes et al. (2006) and Cantley and Cook (2006). For example, how might you make mealtimes more user-focused?

The CSCI has produced a useful bulletin 'Highlight of the Day: improving meals for older people in care homes' Issue 1, 2006 available from: www.csci.gov.uk/professional/default.aspx?page=7327&key

It is recommended that you read the full bulletin, but suggestions include the following:

- **Make food accessible between mealtimes**
  In care homes where food is made accessible between mealtimes and older people are made aware of this, then the embarrassment of asking is avoided. In some care homes, older people feel uncomfortable expressing their preferences directly to staff for fear of being perceived as ungrateful.
- **Use a person-centred planning approach**
  Where person-centred planning is used, care homes are more responsive to changes in a person's care needs. This often involves regularly consulting with the older person, their family members and friends to discover and act on what is important to a person regarding: what foods they can and can't eat; the times they wish to eat; the challenges they face when eating; and the kind of environment in which they like to eat.
- **Supply culturally appropriate food**
  There are cases where the food can be of poor quality and people's cultural needs are not respected. There can be surprising positive benefits for all residents when cultural differences are recognised.
  'A home accommodating one Afro-Caribbean resident did not recognise that people from this cultural background may require culturally appropriate food other then British food. When the home decided to put jerk chicken and other Caribbean food on the menu for all residents, there was a high uptake including from white residents. The home then recognised that all their residents enjoyed food from around the world and satisfied the cultural preferences of the Afro-Caribbean resident.' (Inspector)
- **Use appropriate techniques for serving food**
  Different techniques for delivering food, such as tube feeding, modified textures and adapting crockery and cutlery to maintain independence, are important for people with sensory, physical and cognitive disabilities.
  'There have been times when I have seen a brown splodge on a plate and that has been all the food liquidised together. There are other good providers who liquidise each item separately and present this as an appetising meal, with three or four different portions or colours of food on the plate.' (Inspector)

- **Balance the timing of meals**
  Balancing meals across the day is important because lengthy gaps between meals and fluids can affect many aspects of health, and can lead to sleepiness, lethargy, early waking and poor skin condition.
- **Have a relaxed atmosphere at mealtimes**
  Poor organisation can result in staff focusing on getting through the task at hand. This can unintentionally result in rushing the residents through their meals.
- **Have space to facilitate social interaction**
  It is good practice for care homes to provide a separate room or space for older people who require assistance with eating, to lessen any feelings of embarrassment or loss of dignity. Limited space and inadequate staffing levels for dealing with different eating styles can result in older people eating in their rooms without any real choice in the matter. Inadequate space can lead to problems with eating and can cause the loss of an opportunity for older people to enjoy the sharing of mealtimes with others.

Case study 2 page 110

CASE STUDY

Reviewing service user involvement

Angela runs Brookfield, a residential care unit for people recovering from mental illness. The primary focus of the unit is to provide a 'half-way' support between hospital and fully independent living in the community. Angela has been asked by her manager to review the level of service user involvement in several areas and to make recommendations about which areas should be prioritised and how involvement might be improved. The areas identified are:

5. *Management and staff meetings and forums*
6. *Recruitment and selection of staff*
7. *Training of staff*
8. *Evaluation of the service*

- *Which areas do you think Angela should prioritise and why?*
- *Are there any areas where it would not be appropriate to involve service users and, if so, why?*
- *Should Angela recommend the involvement of individuals currently using services, or individuals who have previously used this or similar services? Why?*
- *Identify key tasks for Angela in beginning to increase the involvement of service users.*

Which areas do you think Angela should prioritise and why?
It may be appropriate to prioritise a more specific area such as recruitment and selection of staff, as this may be more manageable in terms of time and resources. However in the long term it may be most effective to prioritise the involvement of service users in management meetings and forums. This would enable service users to be involved in the meetings or other forums where service user involvement is discussed, and decisions are made about when and how this will take place. Involvement in areas such as training of staff may take longer to initiate as it will involve the need to train service users to deliver training, and to support them in these processes. It may also require the allocation of resources such as finance and staff time to facilitate this which may take more time to plan.

Are there any areas where it would not be appropriate to involve service users and, if so, why?

There will be some meetings and forums where it is not appropriate to involve service users, or only to involve service users for part of the meeting. An example of this would be staff meetings where confidential information relating to other service users may be discussed. There are also likely to be some training courses where it may not be appropriate to involve service users, such as moving, handling or the management of medication.

Should Angela recommend the involvement of individuals currently using services, or individuals who have previously used this or similar services? Why?

Angela may wish to recommend the involvement of both service users who currently use Brookfield's services and those who have previously used services, or use other services. In the short term involving those who have previously used mental health services or use services other than Brookfield might enable Angela to increase service user involvement more quickly. If Angela is able to access a Service User Involvement Project in her area then she may be able to benefit from their expertise and experience in this area, and she may also be able to access training and support services for service users at Brookfield. This may also have the additional benefit of providing activities, support and work to individuals from Brookfield in the future.

Identify key tasks for Angela in beginning to increase the involvement of service users.

Key tasks for Angela might include:

- Establishing whether there are any Service User Involvement Projects in her area, and the type of advice, services and support they might be able to provide.
- Identifying the meetings or forums where it would be most appropriate to involve service users initially. This may include establishing a service user forum if one does not exist. However this does not negate the need to involve service users in other discussion and decision making forums.
- Identifying service users who would be interested in becoming involved and in what capacity.
- Identifying any staff who would be willing to 'champion' and support service user involvement.
- Producing a project plan to discuss with her manager in supervision. Angela may also need or wish to present this to management meetings, staff meetings and service users in order to gain feedback. The project plan should include action points, timescales and resources required at each stage. The plan should also identify how each aspect of service user involvement will be evaluated.

It is important for Angela to identify short and long term plans, and to ensure her short term plans are achievable. This is likely to help staff and service users to see that service user involvement is achievable and the benefits it brings.

Case study 3 page 111

CASE STUDY

Consulting on improvements

Ashford House is a residential home for people with learning difficulties. It also has four semi-independent flats on its site. A recent CSCI inspection has identified that the range and type of activities offered need to be improved to better meet service users' and carers' needs and wishes.

- *What methods would you use to consult with service users and carers?*
- *Would you use the same or different methods to consult with service users and carers? Why?*
- *Why have you chosen these methods? What are their limitations?*
- *How could you ensure that it was not just the views of the more vocal service users and carers that were heard?*

You should consider using a range of methods to consult with service users and carers, and should consider methods which are most appropriate to the needs of those you are consulting.

For example, you will need to consider the level of service users understanding and literacy if you are considering using a questionnaire. You should keep questions short and simple, and may wish to use illustrations alongside questions. You may need to allocate time for staff to support service users to complete questionnaires. It may be a more positive and enjoyable experience to use a method such as graffiti walls, although again you will need to consider individual's ability to contribute and what type of support each individual may need from staff.

While you may feel questionnaires are an appropriate method to use with carers, consider how sure you are that all carers have a level of literacy that would enable them to complete the questionnaire. Also, will you provide a stamped addressed envelope to encourage carers to return the questionnaire to you?

You may also offer some 'surgeries' so those who feel less comfortable in completing forms or questionnaires, or who may simply prefer to have a conversation about their views can participate.

Within focus groups, establishing ground rules from the beginning will help to ensure that all participants contribute within the session. It may be useful to have two members of staff present: one to facilitate the focus group and one to keep notes of the discussion. This will enable the facilitator to focus on the participants and ensuring everyone has an opportunity to contribute to the discussion.

Using a range of methods will also help to ensure that it is not just the views of the more vocal service users and carers which are heard.

Case study 4 page 112

CASE STUDY

Night time care

A study into the night-time care experiences of residents, relatives and staff in three care homes in Scotland (Kerr et al., 2008), highlighted areas in this aspect of provision which could be improved.

Here are some of the problems that were highlighted.

- *The levels of noise and light during the night were too high to support good sleep for residents. The noise was caused by: staff talking; staff activities; buzzers/resident alarm systems; residents; the fabric of the building.*
- *Routine, indiscriminate overchecking led to unnecessary disturbance of residents. One resident said: 'I am fast asleep and then they open the door and put on the light and I jump awake, my heart jumps and then I cannot get back to sleep.'*
- *Night staff were less well trained and less managed and supervised. They sometimes felt isolated and experienced high levels of anxiety – such as about what would happen if there were a fire or if someone needed medical attention.*

- *What improvements might you introduce to overcome these difficulties?*
- *Why?*
- *Who else would you need to involve?*

The recommendations arising from the research are listed below. A summary of the report from which these recommendations are taken, can be found at: http://www.jrf.org.uk/knowledge/findings/socialcare/2201.asp

Recommendations for home management

- Implement regular communication and support strategies between manager and night-time care staff.
- Ensure that environmental concerns within the care home setting are addressed and where appropriate relevant technology is used – for example, guidance around noise, light, safety and silent call systems.
- Put systems in place so that night staff have all the equipment, technology and facilities required to provide good night-time care.
- Monitor staff training requirements, and ensure appropriate times and conditions for such training is provided.
- Keep the use of agency and bank staff to a minimum; where possible, staff with a familiarity of the care setting should be used.

Recommendations for management in connection with night staff

- Implement a system of regular communication with, and supervision of, night-time staff. Give clear messages about the expected night-time practices through specific guidance.
- Ensure a system of training is available to night-time staff and encourage training by ensuring it is night-time specific and at times that do not impact negatively on the night staff. Training content must include dementia awareness.
- Where required, ensure staff are supported to speak English at a level comprehensible to the resident as a basic requirement. Provide basic training where possible, especially where there are difficulties in recruiting night staff.

- Develop and provide guidance to night staff on the impact of night working and strategies to support better health – for example, information on nutrition.

Recommendations for care homes in connection with relatives

- Provide relatives with an information sheet about basic expectations relating to night-time care.
- Include up-to-date photographs of night staff as part of the information.
- Inform relatives of the resident's night-time key worker and encourage some form of regular communication between them.
- Have regular meetings for relatives to improve communication and information sharing.

Recommendations for care homes in connection with residents

- Each resident to have a night-time key worker who will take responsibility for:
    - the production and review of night-time care plans;
    - the communication to other staff of the resident's needs and any changes
    - providing a communication link between the resident and their relatives.
- Night-time care plans should be used to regularly assess and communicate the needs of the resident throughout the night – information should include regular professional assessments of needs such as continence support and pain needs.
- Practices that are intrusive, such as checking and changing pads, should be done with minimal disruption, be gender appropriate and be sensitive to communication needs. They should be in response to individual needs, not part of a group 'round'.

Chapter 7: Safeguarding vulnerable adults

Case study 1 page 117

CASE STUDY

Is this abuse?

Consider the following situations and answer the questions below.

- ***Renuka*** *is a vegetarian because of her religious beliefs. During a stay in a residential home for respite care, the staff ignored this and so Renuka eats very little during her stay.*
- *When they are short staffed, which is often, staff at the* ***Willows Care Home*** *put incontinence pads on the residents so that they do not have to take them to the toilet as often.*
- *The manager at* ***Everdene Residential Home*** *has installed cot sides on most of the beds in the home. She says it helps to stop residents wandering and so 'keeps them safer at night'.*
- *The staff responsible for administering medication at* ***Everdene****, give medication due at night-time at teatime instead so that residents can be put to bed at 7 p.m.*
- ***Sam****, a senior care worker, tells Jim, one of the residents, that if he doesn't 'behave', he will 'have to go' and the next home will be much worse.*
- ***Jean****, a care assistant does some shopping for residents, buying newspapers and small items. She keeps the change because it is too much hassle to do all of the paperwork and the residents 'wouldn't know the difference anyway'.*

In each case, ask yourself these questions.

- *Is this abuse?*
- *If not, why not? If it is, which category of abuse would it fall into?*
- *Does it make any difference whether the abuse is intentional?*
- *If you were aware of this situation, would you report it? Why?*

Each situation is obviously concerning, and it would be important to establish as full a picture of the situation as soon as possible in order to take appropriate action. Even if you feel the abuse is unintentional, the consequences for the service user are likely to be that they have suffered some harm and distress. It does not mean that abuse has not taken place, but your actions may be different, for example if unintentional, you may seek to improve understanding of good practice and to improve performance in the first instance. However, if behaviour continues and is therefore clearly intentional, you will need to take action regarding this.

Renuka

This may be neglect, discriminatory and/or institutional abuse if staff were aware of Renuka's beliefs and dietary needs, but ignored this deliberately. It may also be emotional abuse if this caused distress to Renuka.

Willows Care Home

The needs of staff have taken priority over the needs of the service users and so can be considered institutional abuse. It may also be emotional abuse as service users may feel humiliated or distressed by the actions of staff.

Everdene Residential Home 1

This could be considered physical and/or institutional abuse, because it is inappropriate restraint. The cot sides have been installed to restrict movement, and meets the needs of staff not service users.

Everdene Residential Home 2
As above this could be considered physical and/or institutional abuse. This is misuse of medication, as the medication is given at an inappropriate time to meet the needs of staff not service users.

Sam
This is certainly inappropriate behaviour. In deciding whether this is abuse, and what action you would take you should consider:
a) The consequences to Jim, i.e. the level of distress and harm caused.
b) The intent of Sam, i.e. was she deliberately trying to cause distress, and to threaten Jim?
c) The frequency of this and any similar observations or allegations regarding Sam's behaviour.

Jean
This is financial abuse and a criminal offence as it is theft. It does not matter that the amounts of money which have been kept by Jean may be relatively small.

Signs and Indicators
These lists are **possible** indicators of abuse and the presence of one or more does not necessarily confirm abuse has or is taking place.

Category of abuse	Signs and indicators
Physical	• any injury not fully explained by the history given • injuries inconsistent with the lifestyle of the vulnerable adult • bruises and/or welts on face, lips, mouth, torso, arms, back, buttocks, thighs • clusters of injuries forming regular patterns • burns; friction burns, rope or electric appliance burns • multiple fractures • lacerations or abrasions to mouth, lips, gums, eyes, external genitalia • marks on body, including slap marks, finger marks • injuries at different stages of healing • medication misuse • inappropriate restraint
Psychological/ emotional	• low self-esteem, deference, passivity and resignation • unexplained fear, defensiveness, ambivalence • emotional withdrawal • sleep disturbance • change in, or loss of, appetite
Neglect	• physical condition of person is poor, e.g. bed sores, unwashed, ulcers • clothing in poor condition, e.g. unclean, wet, ragged • inadequate physical environment • inadequate diet • untreated injuries or medical problems • inconsistent or reluctant contact with health or social care agencies • failure to engage in social interaction • malnutrition when not living alone • inadequate heating

Category of abuse	Signs and indicators
Neglect continued	• failure to give prescribed medication • poor personal hygiene • failure to provide access to key services such as health care, dentistry
Sexual	• significant change in sexual behaviour or attitude • pregnancy • wetting or soiling • poor concentration • vulnerable adult appearing withdrawn, depressed, stressed • unusual difficulty in walking or sitting • torn, stained or bloody underclothing • bruises, bleeding, pain or itching in genital area, sexually transmitted diseases, urinary tract or vaginal infection, love bites • bruising to thighs or upper arms • unexplained or unusual responses to personal/medical care tasks
Financial	• unexplained withdrawals from bank accounts • signatures on cheques that do not resemble the person's normal signature, or when the person concerned usually has difficulty writing • inclusion of another person's name on the bank account • numerous unpaid bills when someone is supposed to be paying them on person's behalf • abrupt changes to or the sudden establishment of wills • unexplained transfer of large sums of money or property • person who has previously managed their money well suddenly failing to pay their bills or keep their property maintained as they had in the past • the unexplained disappearance of valuable possessions such as art, silverware, jewellery • someone preventing friends and family from accessing the person, isolating them in order to gain increased control • person becoming anxious and confused about their finances • concern that an excessive amount of money appears to be being spent by care workers on the person's care • lack of amenities (such as TV, toiletries, appropriate clothing) that the person should be able to afford • person being tense after particular people have visited • sudden appearance of previously uninvolved relatives claiming their rights to a person's affairs or possessions
Discriminatory	• lack of respect shown to an individual • failure to respect dietary needs • failure to respect cultural and religious needs • signs of a substandard service offered to an individual • repeated exclusion from rights afforded to citizens such as health, education, employment, criminal justice and civic status

Category of abuse	Signs and indicators
Institutional	• inappropriate or poor care • misuse of medication • inappropriate restraint • sensory deprivation, e.g. denial of use of spectacles, hearing aid, etc. • lack of recording on client files • lack of respect shown to person • denial of visitors or phone calls • restricted access to toilet or bathing facilities • restricted access to appropriate medical or social care • failure to ensure appropriate privacy or personal dignity • lack of flexibility and choice, e.g. mealtimes and bedtimes, choice of food • lack of personal clothing and possessions • lack of privacy • lack of adequate procedures, e.g. for medication, financial management • controlling relationships between staff and service users • poor professional practice • lack of response to complaints

(Adapted from lists produced by Action on Elder Abuse, Age Concern, CSCI and Nottinghamshire Committee for the Protection of Vulnerable Adults)

Case study 2 page 125

CASE STUDY

Consider each of the following situations and identify whether they are a concern, complaint or alert.

The niece of service user **Ethel** *has made an appointment to have a meeting with you. At the meeting, the niece says she is unhappy about the level of cleanliness in her aunt's room, and on a number of occasions has found pieces of old food and dirty underwear on the floor.*

Joyce *has lived in the residential care home you manage for six months. In the past two months, her granddaughter has begun to visit weekly and Joyce is delighted about this. However, staff have informed you that, after a number of visits, Joyce has appeared confused and has told them she has misplaced some money.*

Sami *spent last week having respite care at your setting. The day after he returned home, his mother calls and informs you that Sami has told her he was given meals with meat in them twice last week and he didn't like it. Sami is a vegetarian.*

You have noticed **Ella***'s behaviour has changed recently and she has become more moody and aggressive. You ask her key worker to keep a diary to see if you can identify any patterns or triggers to this, and also to speak to Ella during a calmer moment. The key worker reports that Ella's moods and behaviour coincide with Marvin's shifts and when she asked Ella about this, Ella said Marvin had been showing her 'funny pictures' recently and it made her feel 'strange and not nice'.*

Each situation is obviously concerning, and it is important to establish as full a picture of the situation as possible in order to take appropriate action.

Ethel
This should be considered as a complaint as a specific expression of dissatisfaction has been made by the niece.

Joyce
This situation is concerning and may lead to an alert. The situation would need to be monitored closely, and you may need to review processes for supporting Joyce to manage her finances.

Sami
This is a complaint but may also be an alert, and the situation would need to be explored in more detail. Sami was obviously distressed by this. It would be important to establish why Sami's eating preferences were ignored, e.g. were these recorded fully, were staff aware and if so, why and by who were they ignored. If Sami was given meat as an oversight then it may be decided to treat it as a concern or complaint and resolve it immediately with an apology and ensuring the care plan is followed in future. If staff are offering meat to Sami because they know he does not like it and wish to deliberately cause distress you may well decide that this is an **allegation of abuse** as the vulnerable adult is possibly being psychologically abused.

Ella
This situation is certainly concerning and given the additional information from Ella's key worker should be considered an alert. Your priority must be to safeguard Ella. You should follow your local Safeguarding Adults procedures and speak with a relevant person in the local authority (e.g. Safeguarding Manager) as soon as possible. You should not attempt to speak with, or 'interview', Ella about this situation further, until you have consulted with the local authority.

Chapter 8: Managing budgets

Case study 1 page 136

CASE STUDY

Reviewing the past

Roger has managed Lantern House residential care home for older people since February 2008. One of his first major tasks was to oversee the opening of Lantern House Annexe in April 2008 which allowed Lantern House to increase its provision of respite care.

Roger's manager, Sarah, has asked him to become involved in reviewing and setting budgets for the next financial year (08–09). The homes financial year runs from July to June each year.

Firstly, Sarah asks Roger to review the previous year (07–08) and particularly focus on costs for agency staff; office equipment; gas and electricity and marketing. All of these varied significantly from the projected budget for the year.

- *Look at the Figures 8.1 8.2 and on page 134 and try to identify whether there were particular points in the year when costs peaked or dipped in each category.*
- *What possible reasons are there for these variations from projected costs?*

The costs peaked or dipped in each category for the following reasons:

Agency Staff

The costs of agency staff peak in holiday periods such as August and December which could be anticipated. However Roger may find it useful to review whether it would be possible to 'spread' staff annual leave more evenly across the year. This would make it more manageable for staff to cover for those on leave. There is also a peak of agency staff costs in June which Roger identifies is due to staff 'using up' their leave for the year. This strengthens the argument for the need to spread leave across the year, and is something Roger could discuss in supervision and team meetings.

A peak in costs in February could indicate increased staff sickness which would not be unusual for this time of year. It may be helpful for Roger to look at a number of years figures and see if this is a regular pattern which could be predicted.

Increases in costs in April, May and June also coincide with the opening in the Annexe and an increase in income. This suggests that it would be useful for Roger to look at staffing levels and whether there is a need to recruit more permanent staff.

Marketing Costs

There were no marketing costs for the year which is perhaps unusual considering the opening of the Annexe, and perhaps a need to advertise this increased service. However Roger looks at accounts for 06-07 and finds that a substantial figure was spent at the end of this financial year to cover costs for design and printing of a new brochure for Lantern House. Roger notes that there are enough brochures left to last six months at least, but that some more may need to be printed by the end of the year.

Office Equipment
The projected costs of £50 per month had been allocated to include a warranty for computer equipment and maintenance contract costs for the photocopier. Towards the end of the year the photocopier had begun to break down more regularly. This required an engineer to be called out more often than the number of service visits specified within the maintenance contract, and this had incurred additional premium rate charges.

Gas and Electricity Costs
Roger expected that costs for electricity and gas would increase during the winter months. However during the previous year the cost of gas and electricity had also increased significantly. Roger also notes that costs are likely to have increased due to the opening of the Annexe in April.

Case study 2 page 137

CASE STUDY

Forecasting the future

Having reviewed the past year Sarah asks Roger to look at the same areas (agency staff; office equipment; gas and electricity and marketing) and identify key points for consideration when setting budgets for the following year (09–10). As part of this Sarah asks Roger to identify and review the condition of all office equipment, identifying any which is a priority for replacement in the next year. Sarah also asks Roger to review petty cash expenditure which seemed to increase later in the year.

- *List key questions and areas which you think Roger should consider in relation to the four areas.*
- *Devise a proforma which Roger can use to list all office equipment and justify his priorities for replacement for next year.*

Roger should consider the following issues in the key areas.

Agency staff
- How can annual leave of permanent staff be spread across the year more evenly?
- How could this be monitored and managed more effectively?
- Should more permanent staff be recruited?
- How many?

Marketing Costs
- When are the brochures likely to run out?
- Should more be printed and if so how many?
- What marketing, if any, of the home is required? Consider 'occupation levels' across long stay and respite care.
- Are there more effective and efficient methods of marketing?

Office Equipment
- Would it be cheaper and more effective to replace the photocopier?

Roger notes that the increase in petty cash expenditure seemed to be primarily due to expenditure on printer/photocopier paper. This is a secondary effect of the problems with the photocopier: more photocopier paper had been wasted due to paper jams etc, and staff had begun to rely on the computer printer more heavily as it was quicker and more reliable than using the photocopier.

Roger could use a proforma such as that below to list all office equipment and justify his priorities for replacement for next year.

Item	Age	Current Condition	Expected to last (yrs)	Maintenance costs	Replacement costs	Priority for replacement
Photocopier						

Gas and Electricity Costs

- Are costs likely to increase further?
- Is a cheaper supplier available?
- How much should be built into the budget for the increased costs due to the Annexe?

Case study 3 page 138

CASE STUDY

Setting strategies

Roger has spent considerable time reviewing budgets and considering plans for the next financial year. Most of his discussions have taken place with his manager Sarah.

- *How might Roger have involved staff more fully in the financial reviewing and planning processes?*

How might Roger have involved staff more fully in the financial reviewing and planning processes?

Roger could involve staff in financial reviewing and planning processes through team meetings and supervision. A key area identified from the work Roger has completed so far is staff annual leave. Staff could be asked for possible solutions to help spread this across the year. Staff may also be able to identify possible areas for saving on electricity and gas costs.

Roger could delegate some tasks to other team members, for example the audit of office equipment. Discussion could take place in a team meeting about which items would be priority for replacement.

Petty cash accounts could also be discussed in team meetings. This might help to identify increased costs more quickly.

Case study 4 page 138

CASE STUDY

Setting the budget

You are working with your area manager to set the budgets for the coming financial year. You have collected your financial records from the previous year, which outline all income and expenditure to date. You have a maintenance budget, which has been little used during the last two years. As a result, you have transferred the allocated funds to the utilities budget, where there has been a significant rise in costs, and to the service-user activities budget. You have recently had a health and safety inspection on the home and it has been highlighted that repairs will need to be carried out in the kitchen and one of the bathrooms.

- *How will this affect the way you allocate budgets for the coming year?*
- *How will you ensure there are sufficient funds to pay for the repairs and the increased utility bills?*

Service users have become used to the increase in their activities budget. What implications might the current situation have for service users and staff? If you are going to make changes how will you inform both groups?

How will this affect the way you allocate budgets or the coming year?
All budgets will need to be reviewed to accommodate the rising costs in electric and gas. Look at ways energy can be saved, involving all those who live and work in the setting. The maintenance budget will need to be reinstated and possibly increased.

How will you ensure there are sufficient funds to pay for the repairs and increased utility bills?
As the repairs are essential then budgets will need to be revised. This may mean that other budgets will have to be reduced to accommodate the additional costing.

What implications might the current situation have for service users and staff? If you are going to make the changes how will you inform both groups?
Service users and staff need to be informed of possible changes to budgets as this will have an effect on them – this could be done in staff team meetings and service user meetings. Research if there are grants or other money/services you can access so that service users can still take part in a range of activities.

Case study 5 page 140

CASE STUDY

Delegating financial duties

You have delegated responsibility for food shopping to one of the experienced staff in your team who has said they would like to take on more responsibility. This staff member said they had some experience of this area in a previous job. You did not have time to go through the systems in your setting before he took over this duty, due to staffing shortages. After one month, you check the budget and notice there has been an overspend on the food budget. You have also noticed that food is being thrown away because it has not been used before its sell-by date.

- *Is it the staff member's responsibility that there has been an overspend?*
- *Do you think you prepared the staff member adequately?*
- *What action can be taken to minimise waste and maximise the budget?*
- *How can you support the staff member to develop his skills in this area?*

Is it the staff member's responsibility that there has been an over-spend?
The staff member should not be held responsible as they have only had the responsibility for one month. Knowledge should not be assumed (even if it is an experienced member of staff) and knowledge and experience should have been checked before handing over the responsibility. Do not set staff up to fail by not equipping them with the skills and knowledge they need to complete a task.

Do you think you prepared the staff member adequately?
The staff member was not prepared properly; they were not shown the systems or given time to ask questions or clarify expectations. While they may have had some previous experience all settings/organisations have different ways of doing things. They may not have had responsibility for such a large budget before, which can be quite daunting without guidance and preparation on how to manage it.

What action can be taken to minimise waste and maximise the budget?

There are a number of actions that could be taken to minimise waste and maximise the budget:

- Have weekly menus.
- Involve service users in devising menus to ensure that the food purchased is the food they want to eat.
- Check sell-by dates on food – ensure they are used before they are out of date.

How can you support the staff member to develop his skills in this area?
You can support the staff member by taking the time to show them the systems and letting them shadow you as you complete a food order and balance the budget.

Chapter 9: Interprofessional working

Case study 1 page 150

CASE STUDY

Consider these three scenarios.

Scenario one *A review is scheduled for a service user in your unit. You believe it will take the hour and a half set aside for it, as there are some complex issues for discussion. The social worker calls you to ask if you can 'squeeze in' a review for another service user, as she (the social worker) is behind with reviews and is under pressure from her manager to catch up.*

Scenario two *At your request, the GP comes to visit one of the service users in the unit who is recovering from a minor stroke. The GP says she will arrange for the OT to visit, but when you speak to the OT later in the week, she says that she knows nothing about the service user or referral. The OT also says that, from your description of the issues, it sounds like it is a physiotherapist who is needed not an OT, and that she will not attend.*

Scenario three *You manage a residential unit for people with learning disabilities. Sophie, one of the service users, has been asking for more freedom to go out on her own. The professionals involved all agree that Sophie would be able to do this with an acceptable level of risk and that a plan should be developed to enable this to happen. However, Sophie's parents disagree strongly, saying that Sophie must be accompanied at all times. A meeting is arranged for everyone to attend, including Sophie and her parents, to discuss this further. However, two professionals don't turn up for the meeting and those present don't seem willing to tell Sophie's parents directly that they disagree with their views.*

For each scenario, consider these questions.

- *Why might this be happening?*
- *What strategies can you adopt to deal with the situation while maintaining a positive working relationship with the other professionals?*

Why might this be happening?

Possible reasons are:

- Other professionals may not have the same priorities (in terms of their work load/service priorities) as you.
- Not having the same understanding or perspective of the needs of the service user.
- You not being aware of the roles and responsibilities of other professionals.

What strategies can you adopt to deal with the situation while maintaining a positive working relationship with the other professionals?

Possible reasons are:

- Have realistic expectations of other professionals and their job roles.
- Acknowledge the different aims, work practices and policies of other professionals and their organisations.
- Inform them of your settings, aims, work practices and policies.
- Keep them informed of the needs of the service user in a clear, concise and professional manner.

APPENDIX 2

Useful forms

Gibbs' reflective model template

Description *What happened?*
Feelings *What were you thinking and feeling?*
Evaluation *What was good and bad about the experience?*
Analysis *What sense can you make of the situation?*
Conclusion *What else could you have done?*
Action plan *If it arose again, what would you do?*
This reflective account provides evidence for: (list which units/elements/knowledge statements this will meet)

Completed To-do list

TO-DO LIST

NAME: **DATE STARTED:**

Task to complete	Target date	Tick when completed
Familiarise self with the NVQ units and relate to roles and responsibilities	By the end of next month	
Complete the checklist identifying the purpose of reflection	By the next NVQ workshop	
Complete a reflective account on an experience I have had in the last two weeks using Johns' model of reflection	By the next NVQ W/S	
Complete the same reflective account using Gibbs' model and then identify which model I prefer and why	By the next NVQ W/S	
Complete a reflective account on my job role, identifying my roles and responsibilities. Link this to the NVQ units	By the end of next month	
Practice referencing books and other sources and writing a bibliography I have used when writing my reflective account. Show my NVQ assessor to make sure I have got it right	Next meeting with my assessor	

Completed example of a CPD action plan

What? *What is your CPD need?* *What do you need to learn/ achieve/ research?*	**Why?** *How does this link with your career plan/appraisal/ organisation's business plan?*	**How?** *What type of activity do you need to undertake to achieve the objective?*	**When?** *What is your timescale for completion?*
1. To understand how my team budget works	A key role in my job. Identified in my last appraisal Will help improve financial reporting and identifying problems earlier.	a) To meet with one of the finance officers in the organisation to go through systems b) Read article 'Balancing the books: Budgeting within residential care' Euan Henderson Nursing & Residential Care, Vol. 7, Iss. 2, 19 Jan 2005, pp 87 - 90	a) Within the next month b) Next week
2. To be better at saying 'No'. To feel more confident in being assertive.	It is impacting on my workload and time management Identified in my last appraisal Will increase confidence	To read the relevant chapter in *Personal Effectiveness: A guide to action*, D. Winstanley (2005, CIPD), and complete exercises and try out techniques To review impact with my manager in supervision	a) To read the chapter and complete exercises over next month b) To try out techniques over next 3 months c) To review with my manager in 3 months
3. To chair meetings more effectively	A key role in my job Identified in my last appraisal Would like to improve confidence in this area as part of career development	a) To arrange to shadow 2 colleagues chairing meetings b) To attend a 1-day training course – 17 February. Agreed by manager.	a) Within 2 months b) Within 6 months

Date to be reviewed: ______________________

CPD activity record and learning log

Name: ______________________ **Date:** ____________

Details of CPD Activity
Reading – Kerr, D., Wilkinson, H. and Cunningham, C. (2008) *Supporting older people in care homes at night*, York: Joseph Rowntree Foundation
Learning gained from the activity
Simple measures can have a significant impact, e.g. issues related to light and noise Need to review all areas of service provision, even when there have been no complaints. Many service users in my setting agreed with the issues arising in the research when asked. Important to ensure night-time staff are trained and feel part of the whole team.
How has this learning been applied to/implemented in practice?
Have introduced night-time care plans. Have ensured outstanding basic maintenance, e.g. doors squeaking etc. has been completed. Have revised communication systems.
Did you 'cascade' the information to members of your work team? If yes, how and why; if no, why not?
Presented the research and discussed this in a team meeting. Team agreed to try the use of night-time care plans.
Any future CPD activity or other action required? Identify plan and timescale
Will review the use of night-time care plans with the team and service users in 2 months.

Signature: ______________________ **Date:** ____________

Example of a CPD activity record

Date	**Duration** (hours/days)	**Type of CPD Activity** (reading, shadowing, project work, etc.)	**Details Of CPD Activity**	**Outline your learning from the activity, implications for your practice and any further action required**
21.05.08	7 hours	Conference	'A practical guide to Safeguarding Vulnerable Adults' (Healthcare Events)	1. Provided an update on current policy and thinking around safeguarding. 2. Space to reflect on own attitudes and approaches to this area. 3. Will share information about policy and practice developments with staff in team meeting and develop an action plan for any developments and changes required.
30.07.08	1 hour	Team meeting	Part of team meeting which reviewed the feedback from service users and family members regarding activities currently offered, and suggested changes	1. Good service user and family members response rate. Will use these methods again in the future. 2. Some interesting suggestions for improvements which would not have a major impact on expenditure or staffing needs. This challenged some of my expectations. 3. The team responded positively and are likely to be more comfortable with this type of service user involvement in the future.

A Supported Decision Tool

This tool is designed to guide and record the discussion when a person's choices involve an element of risk. It will be particularly helpful to a person with complex needs or if someone wants to undertake activities that appear particularly risky. (**It can be amended to suit different service user groups)**
It can be completed by the practitioner with the person or by the person themselves with any necessary support, (including the use of communication aids/ pictures where necessary). It is important that, in discussing any risk issues, the person has as much information as possible (in an appropriate form), fully appreciates, and genuinely understands any consequences, to enable them to make their best decisions.
The tool could be adapted for use within existing needs assessment and care planning processes. It also has potential application for any organisation or individual providing advice and support services to people who are self-funders and ineligible for support from their local councils.

Using the tool – Practitioners need to:

- Ensure that the person has the right support to express their wishes and aspirations
- Assume capacity unless otherwise proven
- Consider the physical and mental health of the person and any specialist services they need or are already receiving

Issues for the practitioner to consider
When using the tool with the individual, consider carefully the following aspects of the person's life and wishes:
- dignity
- diversity, race and culture, gender, sexual orientation, age
- religious and spiritual needs
- personal strengths
- ability/willingness to be supported to self care, in terms of:
- opportunities to learn new skills
- support networks
- environment - can it be improved by means of specialist equipment or assistive technology?
- information needs
- communication needs- tool can be adjusted (braille, photo's, simplified language)
- ability to identify own risks
- ability to find solutions
- least restrictive options
- social isolation, inclusion, exclusion
- quality of life outcomes and the risk to independence of 'not doing' .

Supported decision tool

1. What is important to you in your life?	
2. What is working well?	
3. What isn't working so well?	
4. What could make it better?	
5. What things are difficult for you?	
6. Describe how they affect you living your life	
7. What would make things better for you ?	
8. What is stopping you from doing what you want to do?	
9. Do you think there are any risks ?	
10. Could things be done in a different way, which might reduce the risks?	
11. Would you do things differently?	
12.Is the risk present wherever you live?	
13 What do you need to do?	
14. What do staff/organisation need to change?	
15. What could family/carers do?	
16. Who is important to you?	
17. What do people important to you think?	
18. Are there any differences of opinion between you and the people you said are important to you?	
19. What would help to resolve this?	
20. Who might be able to help?	
21. What could we do (practitioner) to support you?	
Agreed next steps-who will do what	
How would you like your care plan to be changed to meet your outcomes?	
Record of any disagreements between people involved	
Date agreed to review how you are managing	
Signature Signature	

APPENDIX 3

Adult Social Care Manager Induction Standards

The Adult Social Care Manager Induction Standards are published by Skills for Care. The following few pages reproduce the standards themselves and the advice given by Skills for Care for using the standards.

Introduction to the standards

Skills for Care has launched the Adult Social Care Manager Induction Standards to make sure new managers hit the ground running as they take up their new posts.

The Adult Social Care Manager Induction Standards are aimed at novice or experienced managers who are new in post in all adult social care settings, regardless of type or size or whether they are in the public or independent sector.

The standards are also designed to be used by people who use services who manage their own directly employed staff, such as personal assistants, and by aspiring managers.

All social care agencies need their leaders and managers to be competent and confident and able to ensure high quality for those who use their services. High-quality induction of new managers is therefore essential. The standards set out clearly for the first time what a new manager needs to know, understand and be able to do.

The basic web edition is free at www.skillsforcare.org.uk, just search on manager induction standards.

Using the standards

The *Options for Excellence* review of the social care workforce, and the White Paper, *Our Health, Our Care, Our Say*, stress the need for all social care agencies to ensure that their leaders and managers are competent and confident to meet their responsibility to provide the highest standards of care.

The Skills for Care leadership and management products provide essential tools for employers and individuals to use in providing this high-quality leadership and management. *Providing Effective Supervision* has received much positive comment since it was launched in July 2007 as part of the latest phase of products and we are now very pleased to launch the *Adult Social Care Manager Induction Standards and Supporting Guidance*.

The social care management induction standards have been developed by Skills for Care to meet an identified gap in the overall range of products that support the national strategy for leadership and management development. They are part of the extended suite of products that support the leadership and management strategy.

The value base

The standards are based upon management practice which has person-centred planning at its heart, with people who use services firmly in control and identifying what is personally important for the achievement of preferred outcomes.

Managers must develop services which are person-centred, seamless and proactive. They should support independence, not dependence, and allow everyone to enjoy a good quality of life, including the ability to contribute fully to their communities. They should treat people with respect and dignity and support them in overcoming barriers to inclusion. Services should be tailored to the religious, cultural and ethnic needs of individuals. They should focus on positive outcomes and well-being, and work proactively to include the most disadvantaged groups.

Managers should fulfil their responsibility to provide care and protection for those who are genuinely unable to express needs or wants or to exercise control. However, the right to self-determination should be at the heart of a reformed social care system and should be constrained only by the realities of finite resources and necessary levels of protection, which should be responsible but not risk averse.

This 'personalised' value base means that the Manager Induction Standards will help social care to meet the seven expected outcomes for people who use adult care services expressed in *Putting People First – a shared vision and commitment to the transformation of adult social care* (DOH 2007), concerning independence, health, freedom from inappropriate social burdens, social and economic participation, quality of life and dignity and respect.

The standards also support the seven 'Common Core Principles to Support Self Care' that promote informed choices, effective communication and confidence, access to information, skills development, use of technology, support networks for care planning and evaluation, and supported risk management, for people caring for themselves. These core principles, published by Skills for Care and Skills for Health in 2008, are a working out of the social model of care.

Who the standards are for

The standards are for all 'new' managers of adult social care – that's those new to management and those new in post, who have previously managed other care services. They are also intended for aspiring or potential managers, to help support their development, although evidence of having met some of the standards will require actual management experience.

The standards are intended to be used in a very wide range of settings – to include people who manage their own services and micro employers as well as small, medium and large organisations across the public, private and voluntary sectors. The standards have been mapped to the core units of the National Occupational Standards for Leadership and Management for Care Services (LMC, first published 2008), referenced in the right-hand column of the web edition of the standards.

New managers should normally have demonstrated all the outcomes within six months of taking up a management role.

Standard 1 Understanding the importance of promoting social care principles and values

Standard 1 is about the key principles and values of social care, which underpin good leadership and management practice.

Main areas	Outcomes	Links to LMC core NOS*
1.1 Principles and values	1.1.1 Know the key principles and values that relate to your work as a manager of social care services	B1.2
	1.1.2 Show how you promote and practise the key principles and values in your day-to-day work as a manager	B1.2
	1.1.3 Know the relevant legislation and regulations that relate to the principles and values of your work	B1.1
	1.1.4 Understand the need to maximise positive outcomes for people who use services	E1.1, E1.2
	1.1.5 Understand the importance of practising ethically, challenging inequality, ensuring inclusion, respecting and promoting diversity and effective anti-discrimination practice	B1.2
	1.1.6 Show how you promote diversity and challenge discrimination in your day-to-day work as a manager	B1.2
1.2 The organisation's values and goals	1.2.1 Understand the values and goals of your organisation or of your management role	B1.1, B1.2
1.3 The multi-agency context	1.3.1 Understand how different organisational cultures and values may impact on partnership working	B1.2
	1.3.2 Know how to minimise or overcome obstacles in partnership working	
1.4 Confidentiality	1.4.1 Know the legal basis for your role as a manager in relation to confidentiality	B1.1
	1.4.2 Understand the importance of the balance between respecting confidentiality and ensuring protection and well-being	C1.2
1.5 Responsibility for the protection of individuals	1.5.1 Know the relevant legislation, policy and procedures	B1.1, B1.3, C1.1, C1.3
	1.5.2 Understand your role and that of others in safeguarding adults	B1.3

* = Leadership and Management for Care Services National Occupational Standards

Standard 2 Providing direction and facilitating change

Standard 2 is about understanding the role of the manager in providing direction and facilitating change within a social and environmental context which includes health and safety.

Main areas	Outcomes	Links to LMC core NOS
2.1 The manager's role	2.1.1 Understand your role, responsibilities and accountabilities as a manager	A1.1
	2.1.2 Understand your organisation's business, workforce and learning plans (as appropriate)	A1.2, A1.4
	2.1.3 Understand your responsibility to maximise positive outcomes for people who use services	E1.1, E1.2
	2.1.4 Show how you understand the changing nature of adult social care and the impact of current developments on your role as a manager	A1.3
	2.1.5 Show how you develop plans for your areas of responsibility	A1.1, B1.1, B1.2, B1.3
	2.1.6 Understand your role in implementing and reviewing plans for your areas of responsibility	
	2.1.7 Show how you manage policies and procedures	B1.1, B1.3, C1.1, C1.2, C1.3
	2.1.8 Know how to lead the engagement and participation of adults and families who use services	B1.1, B1.2, E1.1
	2.1.9 Understand the manager's role in leading and facilitating change	A1.1, A1.2, A1.3
2.2 Risk assessments	2.2.1 Understand how to manage risk	C1.2, C1.3
	2.2.2 Know the principles of risk management and, where appropriate, how to follow risk assessment procedures, including knowing who needs to be informed	C1.2, C1.3
2.3 Health and safety	2.3.1 Understand relevant legislation and regulations and their implications for your work	B1.1, C1.1
	2.3.2 Understand your organisation's health, safety and security policies, systems and procedures and the manager's role in promoting safe working practices (as appropriate)	B1.1, C1.1
	2.3.3 Know how to respond to fire and other emergencies	C1.1, C1.3
2.4 Social context	2.4.1 Understand the context in which your organisation operates	B1.2

Standard 3 Working with people

Standard 3 is about understanding the role of the manager in leading and developing team and individual performance: supervision, recruitment and induction of new staff, ensuring other's learning and effective communication.

Main areas	Outcomes	Links to LMC and core NOS
3.1 The manager's role in supervision	3.1.1 Understand your role in providing effective supervision (as appropriate)	A1.2
	3.1.2 Understand your role in developing productive working relationships	A1.2, E1.1
3.2 Leading teams	3.2.1 Understand your leadership role in a team (as appropriate)	A1.3, A1.4, B1.1, B1.2, B1.3
3.3 Continuing professional development of others	3.3.1 Understand your responsibility for identifying and ensuring the learning of others (as appropriate)	A1.2, A1.4
	3.3.2 Understand recruitment and induction requirements	B1.1, B1.3, E1.3
	3.3.3 Understand the importance of continuing professional development (CPD) in staff retention and workforce development	A1.1, A1.2, A1.4
3.4 Effective communication	3.4.1 Show how you communicate effectively with groups and individuals	E1.1, E1.3
	3.4.2 Understand how information and communications technology (ICT) can help with communication and can effectively share information	E1.2

Standard 4 Using resources

Standard 4 is about the manager's responsibility for using a range of resources: finance, contracts, buildings and technology.

Main areas	Outcomes	Links to LMC and core NOS
4.1 Finance and budgets	4.1.1 Understand your areas of financial accountability and control (as appropriate)	E1.2, E.13
	4.1.2 Understand your responsibility for budgets (as appropriate)	B1.1, E1.3
4.2 Contracts	4.2.1 Understand your responsibilities for supply contracts (as appropriate)	E1.3
	4.2.2 Understand your responsibility for commissioning (as appropriate)	
4.3 Technology	4.3.1 Understand how technology can assist in various elements of your work	A1.3, E1.2
	4.3.2 Show how you use technology effectively in your day-to-day work	A1.3, E1.2
4.4 Buildings and other resources	4.4.1 Understand your responsibility for maintaining buildings and other resources (as appropriate)	B1.1, B1.3, C1.1

Standard 5 Achieving outcomes

Standard 5 is about delivering a quality service that is customer-focused and the manager's responsibilities for performance management processes, partnership working, information sharing, record keeping and change management.

Main areas	Outcomes	Links to LMC and core NOS
5.1 Performance management	5.1.1 Understand your organisation's arrangements for managing performance (as appropriate)	A1.2, A1.3
5.2 Change management	5.2.1 Show how you plan and work for improved outcomes for people who use services	A1.3
	5.2.2 Show how you involve workers and people who use services in service innovation	B1.2, E1.1
5.3 Partnership working	5.3.1 Understand the roles of the partner organisations and the implications for your role	A1.1
5.4 Information responsibilities	5.4.1 Understand the information others may need from you	E1.1, E1.2
	5.4.2 Understand the types of information your team and individual workers may need	E1.1
	5.4.3 Understand the information people who use services may need	E1.1
	5.4.4 Understand the data requirements of the National Minimum Data Set for Social Care and your role (if any) in providing such data	E1.1, E1.2
	5.4.5 Understand feedback and complaints procedures	B1.3, E1.1, E1.3
5.5 Effective recording keeping	5.5.1 Know the importance of managing record keeping	E1.3
	5.5.2 Show how you meet the required record keeping responsibilities	E1.3

Standard 6 Managing self and personal skills

Standard 6 is about managers taking responsibility for their own continuing professional development and the leadership and management skills needed to develop competence in the role.

Main area	Outcome	Links to LMC core NOS
6.1 Continuing professional development	6.1.1 Understand the importance of continuing professional development	A1.1, A1.3, A1.4
	6.1.2 Understand your own learning needs and how they can be met	A1.1
	6.1.3 Show how your day-to-day work has been influenced by feedback from your manager, colleagues, people who use services and carers (as appropriate)	B1.2, E1.1
	6.1.4 Show how you work with your manager or mentor to agree and follow a personal development plan	A1.1
	6.1.5 Understand the methods you can use to improve your work	A1.1, A1.3
6.2 Personal networks	6.2.1 Know the range of people and organisations that can support your work	

certificate of completion

Blank copies of this certificate can be downloaded at **www.skillsforcare.org.uk**

Adult Social Care Manager Induction Standards Certificate of Successful Completion

Name	
Management Role	

I certify that the above named has successfully met all the outcomes in the Adult Social Care Manager Induction Standards. In particular I confirm that:

- An induction plan was agreed and has been followed through to completion.
- Except as below, I have directly assessed the individual's knowledge, skills and understanding and am satisfied that they meet or exceed that required in the standards.
- I have also reviewed any written evidence provided, witnessed or signed off by others and am satisfied with its authenticity and adequacy.
- A continuing personal development plan has been agreed as part of the induction process and there is a written commitment to implement this.
- Other role-specific induction requirements not covered by the standards have been addressed.

Date induction process commenced	
Date Induction Standards completed	
Types of learning included in induction programme	
Signed	
Signatory's name	
Date	
Relationship to inductee	
Organisation/Role (as appropriate)	
Contact address	

"Treat us with respect, support our choices."
"Enable us to access and/or manage the services we need."
"Understand what dignity means in very practical terms and make it a reality."
"Show the way, keep people on board and together."
"Listen to us, make change happen and get results through the best use of people, money and other resources."

What people who use services have told Skills for Care about the social care leaders and managers they want.

what leaders and managers in adult social care do

– a statement for a leadership and management development strategy for social care

There are common behaviours, skills and knowledge required for all adult social care leaders and managers across all sectors and all settings. This statement sets out what is particular about leading and managing in adult social care.

Leaders and managers should work to develop services which are person-centred, seamless and proactive. They should support independence, not dependence, and allow everyone to enjoy a good quality of life, including the ability to contribute fully to their communities. They should treat people with respect and dignity and support them in overcoming barriers to inclusion. Services should be tailored to the cultural, ethnic and religious needs of individuals in the context of their communities. Their leaders and managers should focus on positive outcomes and well-being, and work proactively to include the most disadvantaged groups.

Leaders and managers should fulfil their responsibility to provide care and protection for those who are genuinely unable to express needs or wants or to exercise control. Individuals' self-determination should be at the heart of a reformed social care system and constrained only by the realities of finite resources and by necessary levels of protection, which should be responsible but not risk-averse.

Leadership and management practice needs to be underpinned by the seven *Common Core Principles to Support Self Care.*[1]

Leadership and management practice should also be informed by and rooted in a social model of care.

Leadership and management should be integrated and complementary to each other, so that leadership is reflected in management roles at all levels. Organisations need to develop cultures and structures that will encourage and support leadership and management capability.

Personalised adult social care
All leaders and managers need to ensure their work supports the shared vision as outlined in *Putting People First*[2] and its set of seven outcomes which should ensure that people, irrespective of illness or disability, are supported to:

- live independently
- stay healthy and recover quickly from illness
- exercise maximum control over their own life and, where appropriate, the lives of their family members

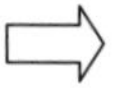

- sustain a family unit which avoids children being required to take on inappropriate caring roles
- participate as active and equal citizens, both economically and socially
- have the best possible quality of life, irrespective of illness or disability
- retain maximum dignity and respect.

Leadership & management resources at www.skillsforcare.org.uk

- Manager Induction Standards – what a manager new in post should know, understand and be able to do. Web edition, plus printed books for new managers and for those who manage or mentor them (order books via www.skillsforcarepublications.org.uk)
- National occupational standards for leadership and management for adult social care – the replacement for the registered manager standards
- *Providing Effective Supervision* – how and why to do it, with real-life models to adapt to your own needs
- Continuing professional development strategy, framework and employer guide – CPD is for everyone, but especially for leaders and managers
- Leadership and management strategy (due late 2008) – an updating of the original 2004 overview and rationale for leadership and management emphasis in social care
- Guide to evaluating leadership & management development (2nd ed, 2006) - how to find out if it is working properly for you

Specifically, leaders and managers need to:

- have a clear vision and be committed to making a positive difference
- work to achieve positive outcomes for people who use services within the context of personalised adult social care
- ensure equality, safety and protection from abuse for staff and people who use services
- address instances of violence against staff and people who use services effectively
- challenge discrimination and harassment in employment practice and service provision
- listen and respond positively to the views of people who use services, carers and staff
- enable staff and people who use services to develop the services people want
- promote and achieve service aims, objectives and goals
- comply with the General Social Care Council's *Codes of Practice*, relevant legislation and agency policies
- develop partnerships and effective joint and integrated working practices
- manage resources and budgets effectively
- manage change effectively
- manage conflicts and risk effectively
- inspire staff
- value people and actively develop talent and potential
- provide effective induction, supervision, performance management and an environment to develop reflective practice, professional skills and the ability to make judgements
- take responsibility for the continuing professional development of themselves and others.

[1] *Common Core Principles to Support Self-Care*, Skills for Care and Skills for Health, 2008, www.skillsforcare.org.uk
[2] *Putting People First*, Department of Health, 2007, www.dh.gov.uk

APPENDIX 4 and Appendix 5

Mapping grids for NVQ/SVQ Level 4 and Management Induction Standards

Appendix 4: Qualification mapping grid for NVQ/SVQ Level 4 Leadership and Management

The mapping grid below gives a comprehensive breakdown of the NVQ/SVQ Level 4 Leadership and Management qualification and indicates the exact location in the book that the information in the units can be located.

Units	Elements	Topic	Where found - chapter and page(s)
A1 Manage and develop yourself and your workforce within the care service	A1.1 Manage and develop self in management and leadership roles	The purpose of reflection and models of reflection	Chapter 1, pages 2–5
		Leadership and management in the care sector	Chapter 2, pages 21–22
		Assessment and identification of CPD needs	Chapter 10, pages 155–6
	A1.2 Manage and develop workers through supervision and performance reviews	Supervision	Chapter 4, pages 59–65
		Appraisals	Chapter 4, pages 65–9
	A1.3 Lead and manage continuous improvements in provision	Identify appropriate learning activities	Chapter 4, pages 68, 71
		Team development	Chapter 5, page 75–9
		Delegation	Chapter 5, pages 89–90
	A1.4 Enhance the quality and safety of the provision through workforce development	Good practice in giving feedback	Chapter 4, pages 64–5
		Team-building events and activities	Chapter 5, pages 77–9

Units	Elements	Topic	Where found - chapter and page(s)
B1 Lead and manage provision of care services that respects, protects and promotes the rights and responsibilities of people	B1.1 Lead and manage prevision that complies with legislation, registration, regulation and inspection requirements	Legal and policy context	Chapter 6, page 102
		Person-centred planning (PCP)	Chapter 3, pages 36–7
		Risk assessment and management	Chapter 3, pages 38–41
	B1.2 Lead and manage provision that promotes the rights responsibilities	Advocacy	Chapter 3, page 38
		Complaints and compliments	Chapter 6, page 107–8
	B1.3 Lead and manage provision that protects people	Legislation and policy that underpins safeguarding vulnerable adults	Chapter 7, pages 118–20
		Categories of abuse; signs and indicators of abuse	Chapter 7, pages 116–18
C1 – Develop and maintain systems, procedures and practice of care services to manage risks and comply with health and safety requirements	C1.1 Implement and monitor compliance with health and safety requirements	Risk assessment and management	Chapter 3, pages 38–41
		Managing poor performance	Chapter 5, pages 90–2
	C1.2 Promote a culture where needs and risks are balanced with healthy and safe practice	Person-centred planning (PCP)	Chapter 3, pages 36–7
		Safeguarding procedures and processes	Chapter 7, page 125
	C1.3 Monitor and review systems, procedures and practice for the management of risk	Evaluating services	Chapter 6, pages 106–10
		Risk assessment and management	Chapter 3, pages 38–41
E1 – Lead and manage effective communication that promotes positive outcomes for people within care services	E1.1 Manage effective communication	Supervision	Chapter 4, pages 59–65
		Factors that help or hinder interprofessional working	Chapter 9, pages 147–50
	E1.2 Ensure that management information systems support the delivery of positive outcomes for people and the provision	Person-centred planning (PCP)	Chapter 3, pages 36–7
		Managing information	Chapter 3, pages 41–3
	E1.3 Manage and maintain recording and reporting systems and procedures and use them effectively	Managing information	Chapter 3, page 41–3

Units	Elements	Topic	Where found - chapter and page(s)
A2 Facilitate and manage change within care services through reflective, motivating and flexible leadership	A2.1 Develop and lead the implementation of a shared vision for your provision	Theories of leadership	Chapter 2, pages 16–8
		Theories of management	Chapter 2, pages 18–9
	A2.3 Develop a culture that is open and facilitates participation	Team development	Chapter 5, pages 75–7
		Leadership and management in the care sector	Chapter 2, pages 21–2
	A2.3 Promote a positive image of your provision and its contribution to the lives of people	Supervision	Chapter 4, pages 59–65
		Change management	Chapter 5, pages 79–81
A3 Actively engage in the safe selection and recruitment of workers and their retention in care services	A3.1 Review the requirements for the safe selection and recruitment of workers, and their retention	Recruitment	Chapter 4, pages 50–2
	A3.2 Actively engage in the safe selection and recruitment of workers	Job description, person specification and application pack	Chapter 4, pages 52–3
		Shortlisting and interviews	Chapter 4, pages 53–4
	A3.3 Implement systems, procedures and practices to support retention	Induction	Chapter 4, pages 57–8
		Induction standards for managers	Chapter 4, page 58
A5 – Allocate and monitor the progress and quality of your work in your area of responsibility (MSC D6)	This unit has no elements	Good practice in giving feedback	Chapter 4, pages 64–5
		Conflict management	Chapter 5, pages 84–6
		Motivation	Chapter 5, pages 86–8

Units	Elements	Topic	Where found - chapter and page(s)
B2 Lead and manage provision of care services that promotes the well being of people	B2.1 Lead and manage provision that involves people in decisions about the outcomes they wish to achieve	Advocacy	Chapter 3, page 38
		Evaluating services	Chapter 6, pages 106–10
	B2.2 Lead and manage provision that promotes people's social emotional, cultural, spiritual and intellectual well being	Leadership and management in the care sector	Chapter 2, pages 21–2
		Appraisals	Chapter 4, pages 65–9
	B2.3 Lead and manage provision that promotes people's health	Learning styles	Chapter 4, pages 70–1
		Supervision	Chapter 4, pages 59–65
B3 Manage provision of care services that deals effectively with transitions and significant life events	B3.1 Implement and review systems, procedures and practice to support people through transitions and significant life events	Erikson's stages of development	Chapter 3, pages 32–3
		Maslow's hierarchy of needs	Chapter 3, pages 33–4
		Kübler-Ross's stages of grief	Chapter 3, pages 34–5
	B3.2 Lead and manage provision that supports people to deal effectively with transitions and significant life events	Coping mechanisms	Chapter 3, pages 35–6
		Change management	Chapter 5, pages 79–82
	B3.3 Implement and review systems, procedures and practice for the sharing of information on transitions and significant life events	Person-centred planning (PCP)	Chapter 3, pages 36–7
		Managing information	Chapter 3, pages 41–3
		Supervision	Chapter 4, pages 59–65

Units	Elements	Topic	Where found - chapter and page(s)
B4 Manage provision of care services that supports parents, families, carers and significant others to achieve positive outcomes	B4.1 Manage effective working relationships with parents, carers, families and significant others	Person-centred planning (PCP)	Chapter 3, pages 36–7
		Managing information	Chapter 3, pages 41–3
		The importance of inter-professional working	Chapter 9, pages 146–7
	B4.2 – Manage systems, procedures and practices to involve parents, carers families and significant others	Legal and policy context of person centred planning	Chapter 3, pages 27–31
		Dignity in care	Chapter 3, page 31
	B4.3 – Support workers to manage situations of conflict	Conflict management	Chapter 5, pages 84–6
B5 Manage and evaluate systems, procedures and practices for assessments, plans and reviews within care services	B5.1 Ensure that workers are competent to carryout assessments, plans and reviews	Legal and policy context of person centred planning	Chapter 3, pages 27–31
		Person-centred planning (PCP)	Chapter 3, pages 36–7
	B5.2 Manage the involvement of people in evaluating the effectiveness of assessments, plans and reviews	Care planning and reviews	Chapter 3, page 38
		Risk assessment and management	Chapter 3, pages 38–41
	B5.3 Evaluate systems, procedures and practices for reviewing the effectiveness of assessments, plans and reviews	Care planning and reviews	Chapter 3, page 38
		Risk assessment and management	Chapter 3, pages 38–41
		Evaluating services	Chapter 6, pages 106–10
D1 Lead and manage work for care services with networks, communities, other professionals and organisations	D1.1 Manage effect working relationships with networks and communities	Person-centred planning (PCP)	Chapter 3, pages 36–37
		The importance of interprofessional working	Chapter 9, pages 146–7
	D1.2 Create and maintain effective working relationships and partnerships with other professionals and organisations	Factors that hinder interprofessional working	Chapter 9, pages 146–9
		Preventative action	Chapter 7, pages 120–125
	D1.3 Contribute to the development of local strategies and services	Professional and personal criteria when working with other professionals	Chapter 9, pages 149–50

Units	Elements	Topic	Where found - chapter and page(s)
E8 Manage finance for your area of responsibility (MSC E2)	This unit has no elements	Budget management	Chapter 8, pages 133–5
		Setting budgets	Chapter 8, pages 135–8
		Monitoring budgets	Chapter 8, pages 138–9
		Financial controls	Chapter 8, page 140

Ten Useful tips for achieving your NVQ/SVQ

There are several useful pieces of information to keep in mind when completing work for your NVQ/SVQ. Some tips for completion of the NVQ/SVQ, including completing the Portfolio can be found below.

- Read through the unit criteria to familiarise yourself with the content of each unit you are going to complete.
- Select your optional units carefully. You may find it helpful to discuss your options with your assessor.
- Meet with your assessor regularly and make sure you have completed any action points they have given you.
- Organise your portfolio early, ensure it is well structured and indexed.
- Plan your study time.
- When writing reflective accounts ensure you give clear and specific examples of your practice. Explain what you have done, why you have done it and what you have learnt as a result.
- Select work product evidence carefully ensuring that it is evident that it is your work and not somebody else's.
- Focus on **quality** of evidence not quantity.
- Include relevant references in your work. You should always reference policy, theory and research.
- There is no such thing as a silly question so when in doubt check it out!

Appendix 5: Qualification mapping grid for Management Induction Standards

The mapping grid below gives a comprehensive breakdown of the Management Induction Standards and indicates the exact location in the book that the information in the units can be located.

Standards	Main Areas	Topic	Where found - chapter and pages
Standard 1: Understanding the importance of promoting social care principles and values	1.1 – Principles and values	Leadership and management in the care sector	Chapter 2, pages 21– 2
		Learning organisations	Chapter 4, pages 46–48
		Legal and policy context of person centred planning	Chapter 3, pages 27–31
		Legislation and policy that underpins safeguarding vulnerable adults	Chapter 7, pages 118–120
		Legislation relevant to recruitment	Chapter 4, page 51–52
	1.2 – The organisation's values and goals	Theories of leadership	Chapter 2, pages 16–18
		Theories on management	Chapter 2, pages 18–19
	1.3 – The multi-agency context	Importance of inter-professional working	Chapter 9, pages 146–147
	1.4 – Confidentiality	Managing information	Chapter 3, pages 41–43
	1.5 – Responsibility to the protection of individuals	Legislation and policy that underpins safeguarding vulnerable adults	Chapter 7, pages 118–120
		Preventative action	Chapter 7, 120–122
		Responding to disclosures or alerts of abuse	Chapter 7, pages 126–128

Standards	Main Areas	Topic	Where found - chapter and pages
Standard 2: Providing direction and facilitating change	2.1 – The manager's role	Leadership/Management	Chapter 2, pages 14–15
		Theories leadership/Management	Chapter 2, pages 16–19
		Person centred planning (PCP)	Chapter 3, pages 36–37
	2.2 – Risk assessments	Risk assessment and management	Chapter 3, pages 38–41
	2.3 – Health and safety	Legislation and policy that underpins safeguarding vulnerable adults	Chapter 7, pages 118–120
		Legislation relevant to recruitment	Chapter 4, page 51–52
		Risk assessment and management	Chapter 3, pages 38–41
	2.4 – Social context	Leadership and management in the care sector	Chapter 2, pages 21–22
Standard 3: Working with people	3.1 – The manager's role in supervision	Supervision	Chapter 4, pages 59–65
	3.2 – Leading teams	Team development	Chapter 5, pages 75–79
		Team effectiveness model	Chapter 5, page 80
	3.3 – Continuing professional development of others	Appraisals	Chapter 4, pages 65–69
		Learning styles	Chapter 4, pages 70–71
	3.4 - Effective communication	Team building events and activities	Chapter 5 pages 77–79
		Supervision	Chapter 4, pages 59–65
Standard 4: Using resources	4.1 – Finance and budgets	Budget management	Chapter 8, pages 133–135
	4.2 – Contracts	Setting Budgets	Chapter 8, pages 135–138
	4.3 – Technology	Selection & Offering the Post	Chapter 4, page 56
		Setting Budgets	Chapter 8, pages 135–138
	4.4 – Buildings and other resources	Setting Budgets	Chapter 8, pages 135–138

Standards	Main Areas	Topic	Where found - chapter and pages
Standard 5: Achieving outcomes	5.1 – Performance management	Appraisals	Chapter 4, pages 65–69
	5.2 – Change management	Change management	Chapter 5, pages 79–82
	5.3 – Partnership working	Importance of inter-professional working	Chapter 9, pages 146–147
	5.4 – Information responsibilities	Managing information	Chapter 3, pages 41–43
	5.5 – Effective record keeping	Recording supervision	Chapter 4, pages 62, 64
		Responding to disclosure	Chapter 7, pages 126–128
Standard 6: Managing self and personal skills	6.1 – Continuing professional development	Benefits of CPD	Chapter 10, page 154
		Strategies for planning and managing your own CPD	Chapter 10, page 154–162
	6.2 – Personal networks	Importance of inter-professional working	Chapter 9, pages 146–147

Bibliography

On the following pages you will find a complete bibliography of all the titles referenced in this book, divided by alphabetical order.

A

ACAS (2006) *Advisory handbook : Discipline and grievances at work*, London: ACAS

Action on Elder Abuse (2006) 'What is Elder Abuse?' www.elderabuse.org.uk (accessed 27.03.08)

ADSS (2005) *Safeguarding Adults: A National Framework of Standards for good practice and outcomes in adult protection work*, www.adss.org.uk/publications/guidance/safeguarding.pdf (accessed 12.07.08)

Age Concern (2008), www.eurolinkage.org/AgeConcern (accessed 07.10.08)

Age Concern Consultation Exercise (2004). Cited in *Better Outcomes for Older People: Framework for Joint Services* (2005), The Scottish Executive, COSLA and NHS

ARC (2005) *My Money Matters! Guidance on best practice in handling the money of people with a learning disabilty*, www.arcuk.org.uk/ (accessed 18.12.07)

Armstrong, M. and Stephens, T. (2005) *A Handbook of Leadership and Management, Assessment Workbook for Assessors* (2nd edition), Leeds: Skills for Care

Atkinson, M., Wilkin, A., Stott, A., Doherty, P. and Kinder, K. (2002) *Multi-agency working: a detailed study*, Slough: NFER

Audit Commission (2004) *Old Virtues, New Virtues: An overview of the changes in social care services over the seven years of Joint Reviews in England 1996–2003*, Wetherby: Audit Commission Publications

Audit Commission, Social Services Inspectorate and The National Assembly for Wales (2004) *Making Ends Meet – Financial Management* from www.joint-reviews.gov.uk/money/Financialmgt (accessed 8.7.08)

Audit Commission (2000) Keeping your balance: standards for financial management in schools. London: Audit Commission

B

Bamber, L. (2007) 'Setting Budgets', *Care Management Matters*, April 2007

Banks, S. (1995) *Ethics and Values in Social Work*, Basingstoke: Palgrave Macmillan

Barrett, G. and Keeping, C. (2005), 'The Processes Required for Interprofessional Working'. In Barrett, G., Sellman, D. and Thomas, J. (eds) (2005) *Interprofessional Working in Health and Social Care: Professional Perspectives*, Basingstoke: Palgrave Macmillan, pp18–31

Barrett, G., Sellman, D. and Thomas, J. (eds) (2005) *Interprofessional Working in Health and Social Care: Professional Perspectives*, Basingstoke: Palgrave Macmillan

Beaty, L. (1997) *Developing your teaching through reflective practice*, Birmingham: SEDA

Belbin, R.M. (1993) *Team Roles at Work*, Oxford: Butterworth Heinemann

Beresford, P. (2005) *Developing Social Care: Service Users Vision for Adult Support*, London: SCIE

Beverly, A. and Coleman, A. (2005) *Induction to Work-based Learning and Assessment Workbase for Assessors* (2nd edition), Leeds: Skills for Care

Beverly, A. and Worsley, A. (2007) *Learning and Teaching in Social Work Practice*, Basingstoke: Palgrave Macmillan

Boud, D., Keogh, R. and Walker, D. (eds) (1985) *Reflection: Turning Experience into Learning*, London: Kogan Page

Brown, A. and Bourne, I. (1996) *The Social Work Supervisor*, Buckingham: OU Press

Burgner, T. (1998) *The Independent Longcare Inquiry*, London: DOH

Burns, R. (1995) *The adult learner at work*, Sydney: Business and Professional

C

Cantley, C. and Cook, M. (2006) *A report on the evaluation of Moor Allerton Care Centre, Northumberland*, Manchester: Dementia North Centre

Chavez, L. and Guido-DiBrito, M. (1999) *New Directions for Adult and Continuing Education*, San Francisco: Jossey Bass

CIPD (2007) *Emotional Intelligence*, London: CIPD

CIPD (2007) *Managing Conflict at Work*, London: CIPD

CIPD (2008a) *Leadership: an overview*, London: CIPD

CIPD (2008b) *What is CPD?* London: CIPD.

Collins, J. (2006) *Good to Great and the Social Sectors*, Colchester: Random House

Cottrell, S. (2003) *The Study Skills Handbook* (2nd edition), Basingstoke: Palgrave

Cree, V. and Wallace, S. (2005) 'Risk and Protection'. In Adams, R., Dominelli, L. and Payne, M. (2005) *Social Work Futures: Crossing boundaries, transforming practice*, Basingstoke: Palgrave Macmillan

Cross, W. (1995) *In Search of Blackness and Afrocentricity: The Psychology of Black Identity Change*, New York: Routledge

CSCI (2005) CSCI (2005) *Care Homes for Older People: national minimum standards*, London: CSCI

CSCI (2006a) *Guidance: Whistleblowing arrangements in regulated care services*, London: CSCI

CSCI (2006b) *In Focus: Quality Issues in Social Care, Issue 4: Safe and sound? Checking the suitability of new care staff in regulated social care services*, London: CSCI

CSCI (2007) *People who use services and experts by experience*, London: CSCI

CSCI (2007a) *Care Homes For Older People: National Minimum Standards*, London: CSCI

CSCI (2007b) *Social Services Performance Assessment Framework Indicators Adults 2006–07*, London: CSCI

CSCI (2007c) *Rights, risks and restraints: An exploration into the use of restraint in the care of older people*, London: CSCI

CSCI (2007d) *In safe keeping: Supporting people who use regulated care services with their finances*, www.csci.org.uk/pdf/in_safe_keeping.pdf (accessed 04.10.08)

CSCI (2008) *Raising Voices: Views on Safeguarding Adults*, London: CSCI

D

Dawson, J. and Barlett, E. (1996) 'Change within interdisciplinary teamwork: one unit's experience', *British Journal of Therapy and Reconciliation* 3:219–22

De Witt, B. and Meyer, R. (2004) *Strategy: Process, Content, Context* (3rd edition), London: International Thompson

DOH (1994) *The Report of the Inquiry into the Care and Treatment of Christopher Clunis*, London, HMSO

DOH (1998) *Modernising Social Services*, London: TSO

DOH (2000) *A Quality Strategy for Social Care*, London: TSO

DOH (2001) *Valuing people: A new strategy for learning disability for the 21st century: planning*

with people: towards person centred approaches – accessible guide
DOH (2001a) *National Service Framework for Older People*, London: DOH
DOH (2001b) *Valuing People: A New Strategy for Learning Disability for the 21st Century*, London: The Stationary Office
DOH (2003a) *National Minimum Standards: Care Homes for Adults (18–65)*, London: TSO
DOH (2003b) *National Minimum Standards: Care Homes for Older People*, London: TSO
DOH (2005) Social Care Green Paper *Independence, Well-being and Choice in Adult Social Care*, London: DOH
DOH (2007a) *Independence, choice and risk: a guide to best practice in supported decision making*, London: DOH
DOH (2007b) *Our Health, Our Care, Our Say*, London: TSO
DOH (2007c) *Social Care Workforce Initiative Newsletter* (June 2007) www.dh.gov.uk/en/SocialCare/Aboutthedirectorate/Researchanddevelopment/index.htm (accessed 15.09.08)
DOH (2008a) *'Direct Payments'* www.dh.gov.uk/en/SocialCare/Socialcarereform/Personalisation/Directpayments/index.htm (accessed 07.09.08)
DOH (2008b) *'Disability'* www.dh.gov.uk/en/SocialCare/Deliveringadultsocialcare/Disability/index.htm (accessed 07.09.08)
DOH (2008c) *Social Care Workforce*, www.dh.gov.uk/en/SocialCare/workforce/index.htm (accessed 15.09.08)
DOH (2008d) *Transforming Social Care*, LAC (DH) (2008) 1
DOH and DFES (2006) *Options for Excellence: Building the Social Care Workforce of the Future*, London: DH/DFES
DOH and Home Office (2000) *No Secrets: Guidance on developing and implementing multi-agency policies and procedures to protect vulnerable adults from abuse*, London: DOH and Home Office

E

Elder-Woodward, J. (2005) *Factsheet 5: Models of Disability*, Edinburgh: UPDATE Scotland's National Disability Information Service

F

Flynn (2007) *The Serious Case Review into the murder of Steven Hoskin* (Flynn, 2007)

G

Gardner, J. (1989) *On Leadership*, New York: Free Press
Gibbs, G. (1988) *Learning By Doing: A Guide to Teaching and Learning Methods*, Oxford: Further Education Unit, Oxford Polytechnic
Goffman, E. (1963) *Asylums: Essays on the social situation of mental patients and other inmates*, Harmondsworth: Penguin
Goleman, D. (1999) *Working with Emotional Intelligence*, London: Bloomsbury
GSCC (2002) *Code of Practice for Social Care Workers and Code of Practice for Employers of Social Care Workers*, London: GSCC
GSCC (2007) *Guidance notes on how to renew your registration*, London: GSCC
GSCC (no date) *Post-Registration Training and Learning (PRTL) requirements*, London: GSCC

H

Hawthorne, L. (1975) 'Games Supervisors Play', *Social Work*, 20(3):179–183
Healthcare Commission (2007) *Investigation into the service for people with learning disabilities provided by Sutton and Merton Primary Care Trust*, www.healthcarecommission.org.uk/_db/_documents/Sutton_and_Merton_inv_2006_easyread.pdf (accessed 02.05.08)
Healthcare Commission and CSCI (2006) *Joint Investigation into the provision of services for people with learning disabilities at Cornwall Partnership NHS Trust*, www.healthcarecommission.org.uk/_

db/_documents/cornwall_investigation_report.pdf (accessed 07.06.08)

Herzberg, F., Mausner, B. and Snyderman, B.B. (1959) *The Motivation to Work*, New York: John Wiley

Holman, D., Pavlica, K. and Thorpe, R. (1997) 'Rethinking Kolb's theory of experiential learning in management education: the construction of social constructionism and activity theory', *Management Learning* 28(2):135–148

Holmes, J. (2004) *John Bowlby and Attachment Theory*, Hove: Brunner-Routledge

Honey, P. and Mumford, A. (1986) *Manual of Learning Styles*, Maidenhead: Peter Honey

HPC (2004) *Continuing Professional Development – Consultation Paper*, London: HPC HPC (2006) *Continued Professional Development and Your Registration*, London: HPC

HSE (2008) *Management Standards for Stress: Step 3* www.hse.gov.uk/stress/standards/step3/index.htm (accessed 03.04.08)

HSE, Department for Work and Pensions, Cabinet Office and The Work Foundation (2006) *Ministerial Taskforce on Health, Safety & Productivity: The Well Managed Organisation Diagnostic Tools For Handling Sickness Absence*, www.hse.gov.uk/services/pdfs/diagnostictools.pdf (accessed 03.04.08)

Hunt, J.W. (1992) *Managing People at Work* (3rd Edition), Berkshire: McGraw-Hill

I

Improvement and Development Agency and Audit Commission (2006) *Managing Quality* www.makingendsmeet.idea.gov.uk

Innes, A., Macpherson, S. and Mccabe, L. (2006) *Promoting person-centred care at the front line*, York: Joseph Rowntree Foundation/SCIE

International Stress Management Association, HSE & ACAS (2004) *Working together to reduce stress at work: A guide for employees*, International Stress Management Association

J

Johns, C. (1994) 'Guided reflection'. In A. Palmer, S. Burns and C. Bulman (eds) (1994) *Reflective Practice in Nursing: The growth of the professional practitioner*, Oxford: Blackwell Science

Kadushin, A. (1968) 'Games people play in supervision', *Social Work*, 73:127–136

Kayes, D.C. (2002) 'Experiential Learning and Its Critics: Preserving the Role of Experience in Management Learning and Education', *Academy of Management Learning and Education*, 1(2): 137–149

K

Kerr, D., Wilkinson, H. and Cunningham, C. (2008) 'Supporting older people in care homes at night', York: Joseph Rowntree Foundation

Kolb, D. (1984) *Experiential Learning: Experience as a Source of Learning and Development*, New Jersey: Prentice Hall

Krathwohl, D., Bloom, B. and Masia, B. (1964) *Taxonomy of Educational Objectives: The Classification of Educational Goals Handbook II: Affective Domain*, New York: David McKay

Kroger, J. (1997) 'Gender and identity: The intersection of structure, content, and context', *Sex Roles* 36 (11/12): 747–770

Krüger, W. (1996) 'Implementation: The Core Task of Change Management', *CEMS Business Review*, Vol. 1, 1996

L

Laird, D. (1985) *Approaches to training and development*, Reading, Mass: Addison-Wesley

Lord Laming (2003) *The Victoria Climbié Inquiry: Summary Report of an Inquiry*, Cheltenham: HMSO

Lee-Treweek, G. (1998) 'Bedroom Abuse: the Hidden Work in a Nursing Home'. In Allott, M. and Robb, M. (eds) (1998) *Understanding Health and Social Care: An Introductory Reader*, London: Sage

Leicester City Council (2004) *Adult Protection: Preventing Abuse of Vulnerable Adults* (2nd edition)

Lewin, K., Lippit, R. and White, R.K. (1939) 'Patterns of aggressive behavior in experimentally created social climates', *Journal of Social Psychology*, 10: 271–301 London: Kogan Page

Livy, B. (1988) *Corporate Personnel Management*, London: Pitman

London Councils (no date) *Workforce Analysis and Planning guide*, www.londoncouncils.gov.uk/improvement/hrip/workforceanalysis/default.htm (accessed 27.09.08)

Love, C. (1997) *Developing People The Manager's Role – 20 Tried and Tested Activities for Helping Managers Foster a Positive Learning Climate in the Workplace*, Ely: Fenman

M

Manzoni, J. and Barsoux, J. (2002) *The Set-Up-to-Fail Syndrome: How Good Managers Cause Great People to Fail*, Boston: HBS Press Book

Mason, L. (2003) 'Leading Teams in the 21st Century'. In Thomas, A. (2003) *Leading and Inspiring Teams* Oxford: Heinemann Educational

McClelland, D. (1988) *Human Motivation* Cambridge, Cambridge University Press

McGill, I. and Brockbank, A. (1998) *Facilitating reflective learning in higher education*, Buckingham: Open University

Means, R., Richards, S and Smith, R (eds) (2003), *Community Care: Policy and Practice* (3rd edition), Basingstoke: Palgrave Macmillan

Merriam, S. and Caffarella, R. (1999) *Learning in Adulthood: A Comprehensive Guide* (2nd edition), San Francisco: Jossey-Bass

Mind (2008) *The Mind Guide to Advocacy*, London: Mind

Molyneux, J. (2001), 'Interprofessional Teamworking: what makes a team work well?', *Journal of Interprofessional Care* 15:29–35. Cited in Barrett, G. and Keeping, C. (2005) *The Processes Required for Effective Interprofessional Working*

Moon, J. (1999) *Learning Journals: A Handbook for Academics, Students and Professional Development*, London: Kogan Page

Moon, J. (2001) 'The development of assessment criteria for a journal for PGCE students' (unpublished) University of Exeter. In Watton, P., Collings, J. and Moon, J. (2001) *Reflective Writing: Guidance notes for students*, www.ex.ac.uk/employability/students/reflective.rtf (accessed 26.08.08)

Moriarty, J. (2005), 'The future of social care', *Journal of Dementia Care* 13(3):10–11

Mukherjee, S., Beresford, B. and Sloper, T. (1999) *Unlocking Key Working: An Analysis of Keyworker Services for Families with Disabled Children*, Bristol: The Policy Press

N

NHS Leadership Centre (2003) *NHS Leadership Qualities Framework* www.nhsleadershipqualities.nhs.uk (accessed 03.09.08)

Nottinghamshire Committee for the Protection of Vulnerable Adults (2007) *Nottingham and Nottinghamshire Safeguarding Adults Multi-Agency Procedure*

O

O'Rourke, M. (1999) 'Dangerousness: how best to manage the risk', *The Therapist* 6(2)

Oliver, M. (1990) *The Politics of Disablement*, New York: St Martin's Press

P

Penhale, B., Perkins, N., Pinkney, L., Reid, D., Hussein, S and Manthorpe, J. (2007) *Partnership and regulation in adult protection: The effectiveness of multi-agency working and the regulatory framework in Adult Protection*, London: DOH, Social Care Workforce Unit and The University of Sheffield

Platt, D. (2003) 'Surviving and Thriving in a Changing World.' Speech made at NATOPPS conference 28.4.03 www.dh.gov.uk/en/News/

Speeches/Speecheslist/DH_4031709 (accessed 12.09.08)

Platzer, H., Blake, D. and Snelling, J (1997) 'A Review of Research into the Use of Groups and Discussion to Promote Reflective Practice in Nursing Research'. In *Post Compulsory Education*, 2(2):193–204

Pring, J. (2005) 'Why it took so long to expose the abusive regime at Longcare'

Public Concern at Work (no date) *Best Practice Guide*, London: Public Concern at Work

Q

Quinney, A. (2006) *Collaborative Social Work Practice*, Exeter: Learning Matters

R

Rabiee, F (2004) 'Focus-group interviews and data analysis: Proceedings of the Nutritional Society' 63:655-660

Randall, G. (1989) 'Employee Appraisal'. In Sisson, K. (ed) *Personnel Management in Britain*, Oxford: Basil Blackwell

Reder, P., Duncan, S., Gray, M. and Stevenson, O. (1993) *Beyond Blame: Child Abuse Tragedies Revisited*, Abingdon: Routledge

Roberts, D., Scharf, T., Bernard, M., and Crome, P. (2007) *Identification of deafblind*
dual sensory impairment in older people, London: SCIE

Rothwell, A. and Arnold, J. (2005) 'How HR professionals rate continuing professional development', *Human Resource Management Journal*, Vol 15, No. 3:18–32. Cited in Megginson, D. and Whitaker, V. (2007) *Continuing Professional Development* (2nd edition), London: The Chartered Institute of Personnel and Development

Rummery, K. (2002) 'Disability, Citizenship and Community Care: A Case for Welfare Rights'. Cited in Means, R., Richards, S. and Smith, R. (eds) (2003) *Community Care: Policy and Practice* (3rd edition), Basingstoke: Palgrave Macmillan

S

SCIE (2004) *Learning Organisations: A self-assessment resource pack*, www.scie.org.uk/publications/learningorgs/index.asp (accessed 27.09.08)

SCIE (2004) SCIE position paper 3: Has service user participation made a difference to social care services? London: SCIE

SCIE (2005) *Performance Appraisal in a Nutshell*, www.scie-peoplemanagement.org.uk/resource/docPreview.asp?surround=true&lang=1&docID=128 (accessed 27.09.08)

SCIE (2006) *Common Induction Standards: Guidance for those responsible for workers in an induction period*, Leeds: Skills for Care

SCIE (2007) *Workforce Planning: A mini guide*, www.scie-peoplemanagement.org.uk/resource/docPreview.asp?surround=true&lang=1&docID=137 (accessed 27.09.08)

SCIE (2008) *People Management* www.scie.org.uk/workforce/peoplemanagement.asp (accessed 03.05.08)

SCIE (no date) Practice *Guide 1: Managing Practice*
www.scie.org.uk/publications/practiceguides/bpg1/index.asp (accessed 27.09.08)

Scott, C. and Jaffe, D. (2004) *Managing Change at Work: Leading people through organizational transitions* (3rd edition), Boston: Thompson Learning

Scottish Commission for the Regulation of Care (2004) *A review of the quality of care homes in Scotland 2004* www.carecommission.com/images/stories/documents/publications/reviewsofqualitycare/197.pdf (accessed 27.09.08)

Scottish Executive (2004) *Investigations into Scottish Borders Council and NHS Borders Services for People with Learning Disabilities: Joint Statement from the Mental Welfare Commission and the Social Work Services Inspectorate*

Scottish Executive (2005a) *Better Outcomes for Older People*, Edinburgh: Scottish Executive

Scottish Executive (2005b) *National Strategy for the Development of the Social Service Workforce in Scotland: A Plan for Action 2005–2010*, Edinburgh: Scottish Executive

Scragg, T. (1995) *Managing at the Front Line: A handbook for managers in social care agencies*, Brighton: Pavilion

Skills for Care (2006) *Leadership & Management Strategy: A Strategy for the Social Care Workforce*, Leeds: Skills for Care

Skills for Care and CWDC (2007) *Providing Effective Supervision*, Leeds: Skills for Care/ CWDC

Sloper, P. (2004) 'Facilitators and barriers for co-ordinated multi-agency services', *Child Care, Health and Development*, 30(6):571–580

Stewart, R. (1997) (3rd ed) *The Reality of Management*, Oxford: Butterworth-Heinemann

Stogdill, R.M. (1974) *Handbook of Leadership: A survey of the literature*, New York: Free Press

Stoke-on-Trent City Council (2008) Adult Social Care Quality Standards. www.stoke.gov.uk/ccm/ content/ss/adult-social-care-quality-service-standards.en (accessed 14.08.08)

Strebler, M. (2004) *Tackling Poor Performance* (Report 406), Brighton: Institute for Employment Studies

T

The Alzheimer's Society (2008) *What Standards of Care Can People Expect From a Care Home?*, www. alzheimers.org.uk/ (accessed 04.06.08)

The Circles Network (2008) *What is Person Centred Planning?* , www.circlesnetwork.org.uk/what_is_ person_centred_planning.htm (accessed 04.06.08)

Thomas, K.W. and Kilmann, R.H. (1974) *Thomas Kilmann Conflict Mode Instrument*, California: CPP Inc.

Thompson, N. (2006) *Promoting Workplace Learning*, BASW: Policy Press

Titterton, M. (2005) *Risk and Risk Taking in Health and Social Welfare* London: Jessica Kingsley

TNS UK Ltd (2007), *National Survey of Care Workers: Final Report*, Leeds: Skills for Care

Topss (2004) *Workforce Planning Toolkit* (2nd edition), Leeds: Topss

Tuckman, Bruce W. and Jensen, Mary Ann C. (1977) 'Stages of small group development revisited', *Group and Organizational Studies*, 2: 419–427

U

University of Victoria (2008), *Team Effectiveness Model* web.uvic.ca/hr/hrhandbook/organizdev/ teammodel.pdf (accessed 03.05.08)

V

Van den Hende, R. (2001) 'Public concern at work: supporting public-interest whistle blowing', *Journal of Adult Protection* 3(3):41–44. Cited at SCIE Practice Guide 09: Dignity in care, 2008, www.scie.org.uk/publications/practiceguides/ practiceguide09/files/pg09.pdf (accessed 23.09.08)

W

Wardhaugh and Wilding (1998) 'Towards an explanation of the corruption of care', *Critical Social Policy*, 3: 4–31

Watson, Joan E.R. (2007) '"The Times They Are A Changing" – Post Qualifying Training Needs of Social Work Managers', *Social Work Education*, 27(3):318–333

Wilderom, C.P.M. (1991) 'Service management/ leadership: different from management/leadership in industrial organisations?', *International Journal of Service Industry Management*, vol 2(1):6–14

Winter, R., Buck, A. and Sobiechowska, P. (1999) *Professional Experience and the Investigative Imagination*, London: Routledge

Index